Richard Howells studied law at the University of Wales and the London School of Economics. He has experience of occupational safety in the coal industry and as a factory inspector. He has taught law for a number of years and is now Professor of Law at the Polytechnic of Central London's School of Law. His research and publications have been on the interpretation and enforcement of safety law and labour legislation.

Brenda Barrett studied law at Oxford and was called to the Bar at Gray's Inn in 1959. After some practical experience in industry and at the Bar, she entered the academic world. She now runs the law degree course at Middlesex Polytechnic. She has long experience of lecturing and publishing in the field of labour law. She has also undertaken research into management and employee attitudes in respect of safety.

The authors have co-operated in occupational safety research into harmonization of laws in the EEC and also into offshore installations.

Contents

1 Background to the Present Law

Introduction

This book sets out to provide a guide for management regarding the law relating to health and safety at work, with particular emphasis upon the Health and Safety at Work etc Act 1974 (the Act). This Act received the Royal Assent on 31 July 1974 and became fully operative on 1 April 1975. It is now the major legislative provision relating to occupational health and safety. The objectives of the Act, which are very ambitious, include both raising the standards of safety and health for all persons at work, and the protection of the public, whose safety and health may be put at risk by the activities of persons at work. Because it is of general application, it brings within statutory protection many classes of persons (often referred to as 'new entrants') who were previously unprotected.

The main aim of the Act is to raise safety standards generally; however, its initial impact has been on the ways in which work is organized, supervised and carried out. The effects of the Act are therefore being felt by the general manager and the personnel manager.

The book takes the form of a commentary for management upon their responsibilities under the Act and does not attempt to paraphrase the actual provisions of the Act. The Act is printed in full in appendix I on page 149. The reader should bear in mind that only the statutory text is authoritative, and that the commentary will need constant modification in the light of regulations which will be issued from time to time, reported decisions in which the Act has been interpreted by the courts, and the spirit in which its new requirements are administered by the enforcement agencies. He should also remember that, for the time being, the law enacted in this Act must be read in conjunction with much of the earlier statute law and case law relating to occupational health and safety.

The general scheme of the work is a series of chapters, divided into short sections dealing with specific topics, combining commentary with references to legal requirements. As far as possible the sections follow the same order as the provisions of the Act. Part III of the Act deals with the somewhat distantly related subject of building construction standards. For this reason Part III is not discussed in the text, nor are its provisions set out in appendix I on page 149.

The historical development of safety legislation

Legislation laying down minimum standards of safety, health and welfare originated in the United Kingdom more than 100 years ago. The subsequent development of the legislation has been remarkably piecemeal. By 1970 approximately two-thirds of the workforce of this country were protected by a patchwork of separate codes of law made up of more than 30 statutes and nearly 500 statutory instruments.

Much of this legislation followed a common pattern, which had survived more or less unaltered in essentials from the early enactments, and which had the following characteristics:

(a) a basic statutory code, supplemented and updated by regulations, relating primarily to technical, engineering and scientific matters concerned with the physical environment of the workplace
(b) enforcement by a specialist inspectorate, with statutory powers to enter premises, interrogate persons and prosecute offenders
(c) reliance upon petty criminal sanctions of a financial nature for punishment of offenders
(d) the imposition of general responsibility upon employers with residual disciplinary duties upon employees
(e) legal responsibilities usually imposed upon organizations and not normally upon managers or officials.

A weakness of this statutory pattern was the emphasis it placed on the control of employment conditions by means of duties imposed upon employers in respect of their own premises. This emphasis resulted in a failure to protect either 'visiting workers' or those workers or other persons who were 'neighbours'[1] of industrial activities.

The original nineteenth century decision to impose criminal sanctions on employers for enforcement purposes has had a far reaching effect upon the development of safety legislation. As the scale of industrial activity increased, and the ownership of premises and employment of labour passed into the hands of corporations, the criminal liability placed on the 'employer' was in effect liability for functions delegated by him to his managers. The difficulty of reconciling the emphasis, under this legislation, on the employer's responsibility for the faults of his managers, with the need for proof of personal guilt normally required by the criminal law, led to an uneasy compromise in the twentieth century legislation. A heavy reliance was placed upon offences of strict liability which were punished lightly; the penalties being almost exclusively financial.

Mining legislation did not follow this pattern. In this instance the policy, which can be seen in the legislation of 1954, was to regulate the system of management as well as regulating technical mining standards. It imposed criminal liability directly upon members of the management chain, such as managers, under-managers and officials of mines, as well as upon owners[2]. Perhaps in consequence, heavier penalties, including imprisonment were laid down for offenders against the mining safety code.

In addition, factory and mine inspectors had over the years become responsible for the enforcement of much unrelated legislation (eg the Truck Acts and the Employers' Liability (Compulsory Insurance) Act).

It is important to bear the basic characteristics of the previous safety legislation in mind when approaching the Act, as it represents a complete break with the previous pattern.

The Robens Committee and Report

In the 1960s there were signs of increasing dissatisfaction with the results being achieved by the traditional safety legislation, although it was recognized that it had been successful in bringing about a long-term reduction in the levels of accidents and diseases. A Bill was introduced into Parliament seeking to provide for compulsory worker involvement in accident prevention. The Bill, although not enacted, focused Parliamentary attention on the accident problem. The result was that a Committee on Safety and Health at Work, under the chairmanship of Lord Robens, was set up by Barbara Castle in May 1970. The terms of reference of the Committee were the widest possible, relating to the whole field of occupational risks and extending to the protection of the public from hazards arising in connection with work activities. The only workers excluded from this enquiry were transport workers[3] whilst directly engaged in transport operations, since they were protected by other legal provisions.

The Robens criticisms

The Report[4] itself is an interesting document, being the first attempt to survey the whole problem of occupational health and safety in the United Kingdom in order to identify the defects in the traditional approaches to the problem and to make recommendations for a completely new system of legislation to achieve its objectives. An appreciation of this Report is essential to an understanding of the subsequent legislation, in which Parliament has attempted to implement both the general philosophy and the detailed recommendations of the Robens Report.

The Report contains a series of fundamental criticisms of the then existing legal framework, linked with suggested remedies. These include:

(a) the empirical approach of the law to a problem which needed to be approached in a systematic manner. To remedy this the mass of legislation should be replaced by a single statute of general application
(b) the complexity and superfluity of the law. To remedy this the mass of detail should give way to a few simple, easily assimilated precepts
(c) the lack of enforcement procedures. To remedy this prosecution should not be a matter of first resort, but the remedies available should be adequate to ensure compliance in all cases
(d) the exclusive emphasis upon the workplace environment. Occupational safety legislation should also protect visitors and the public in general
(e) the undue emphasis upon technical safety standards and neglect of systems of working. This imbalance should be remedied by placing greater emphasis upon management's responsibilities for providing and maintaining safe systems of working

(f) the failure to involve the workforce in the safety effort. Nothing but good could result from the active involvement of workers in the procedures for accident prevention at their place of work.

The Robens proposals

To remedy these shortcomings and carry out the recommendations the Report proposed the formulation of a new philosophy for accident prevention. It contended that there were severe practical limits to the extent to which progressively better standards of safety and health at work could be brought about through negative regulation on the lines previously adopted. That system had concentrated rather too much on external regulation and too little on personal responsibility and voluntary effort. The Committee believed that the primary responsibility for doing something about the then-prevalent level of occupational accidents and diseases, lay with those who created the risks, and with those who worked with them. Although this responsibility lay upon both workers and employers, management was in the best position to give a lead in safety. Indeed, promotion of safety and health at work was an essential function of good management, requiring positive attitudes to safety from directors and senior managers.

The new role of the law

It was unfortunate that the Robens emphasis upon safety, as a normal management function, tended to create the impression that a voluntary, as opposed to a legally enforced, system was favoured by the Committee. In reality, the detailed proposals it made called for a legal machinery designed to motivate employers and employees to go beyond the minimum legal requirements of a statutory code. Such a system was intended to create a climate of opinion in which safety would be seen as a normal business objective; thus the enforcement of the basic statutory duties should be supplemented by the voluntary effort of all concerned.

The Committee was clearly not convinced that safety and health at work could be ensured by an ever expanding body of mandatory standards, but it was equally convinced that the role of law remained crucial; and that this law should be simple, of general application and firmly enforced.

The natural implications of many of the Committee's recommendations involved wider powers for inspectors and greater responsibilities for employers and managers than anything that had previously been seen. A necessary consequence of those recommendations was the introduction of an element of State control into the day to day management of enterprises, to an extent beyond that found in previous safety legislation.

Implementation of the Report

In May 1973 the Government stated its intention of introducing legislation implementing the main recommendations of the Robens Committee. Legislation was introduced in 1974, and survived the change of Government in that year, to become law in substantially unaltered form.

In the following chapters we shall be examining the provisions in which these recommendations were implemented. It will be appreciated that we are dealing with what is known as 'enabling legislation'. For this reason its provisions are of necessity wide, and remain somewhat vague, except where they have been interpreted by the courts or augmented by regulations and codes of practice. So the Robens Report, upon whose recommendations the Act is so clearly modelled, will remain valuable as a guide to the intention of the Act and will, upon occasion, suggest the ways in which the wide powers it contains are likely to be administered in practice; even though, with the passage of time, the Robens philosophy itself may no longer be accepted to the same extent.

2 The Scope of the Health and Safety at Work Act

Objectives of the Act

(Section 1 and Schedules 1, 3)
The 1974 Act takes the form of an enabling statute. That is, the Act provides legal machinery for the amendment of the earlier law so as to achieve certain stated objectives. Thus an understanding of the Act requires a grasp of these objectives. These are generally to provide a framework within which the protection of safety legislation is extended to all persons at work in the United Kingdom, and also to persons other than those at work, against the risks created by the activities of persons at work. At the same time earlier legislation and regulations have remained in force until replaced by equivalent (or more effective) requirements forming part of the new legislative code. The term 'existing statutory provisions' is used in the Act to describe this earlier legislation (listed in Schedule 1,) together with the regulations in force thereunder, and the term 'relevant statutory provisions' is used to describe the complete legislative code in force under this legislation, including the 1974 Act and regulations[1].

The general purposes of the Act are set out in s1. The section is of fundamental importance because it spells out the objectives of the legislature and thus gives essential practical guidance to those who have to work with the Act, in particular employers, employees and the enforcement agencies. The judges, who are required to interpret the provisions of the Act, may also find this guidance as to its general objectives a source of assistance.

Section 1(1) states that the provisions of the Act shall have effect with a view to securing the health, safety and welfare of persons at work and protecting persons other than those at work, against risks to health and safety arising out of, or in connection with, the activities of persons at work. The purposes could hardly be expressed more widely and thus demonstrate the sharp contrast between this Act and previous legislation relating to occupational health and safety. In the past legislation had tended to be restricted to categories of people at work in specific classes of premises or engaged in specific activities. The scope of this section is not confined or restricted in this way – if there is a situation in which persons are at work[2], regardless of the legal relationship under which they are employed, or their status, or the activity they are engaged upon – the Act applies[3].

The extension of the objectives of occupational safety legislation to cover the protection of the public was also a novel development. Occupational safety legislation in the past had been enacted solely for the protection of categories of persons at work, and had stopped short of empowering the inspectorates to enforce such legislation with a view to the protection of the public: nor did it confer power to make valid regulations for the safety of the public. However, in the 1960s and early 1970s industrial disasters endangering the public, such as the Brent Cross coach disaster 1964, Aberfan 1966 and Flixborough 1974 served to emphasize the unsatisfactory nature of the hiatus in the then existing law.

Section 1(1) makes it clear that, for the future, hazards will be regulated for the protection of the public as well as for the protection of persons at work.

Thus the inspectorate is able to exercise its enforcement powers with the protection of the public in mind, and regulations and codes of practice may be made with the specific purpose of protecting the public.

A further purpose of the subsection is to control the keeping and use of explosives, highly flammable and other dangerous substances. Work with these substances is, of course, governed by the general requirements of the Act; so it follows that an object of the Act is to regulate storage of these substances whether or not persons work on the premises in question.

Lastly, the emission of noxious or offensive substances from prescribed premises into the atmosphere is now controlled under this legislation. At first sight the connection between this subject and occupational safety and health may seem tenuous, but given that the protection of the public is an object of the Act, its relevance becomes clear.

Section 1(2) sets out the general objectives of the regulation-making powers of the Act. Regulation making is examined in detail in chapter 6 but must also be considered in this chapter because it throws light on the objectives of the Act.

The subsection creates wide powers and duties concerning the amendment by means of regulations of existing occupational health and safety legislation, and the making new regulations and codes of practice for the purpose of achieving the general objects of the Act[4].

As the earlier statutory provisions are reviewed, s1(2) requires their replacement by a system of regulations and approved codes of practice designed to maintain or improve the standards of health, safety and welfare established by or under the earlier enactments.

The extent of these regulation-making powers will be apparent if these objectives are read together with Schedule 3, which sets out the subject matter of health and safety regulations.

It will be appreciated that the earlier statutory provisions varied greatly in the stringency of their requirements and the effectiveness of their legal and administrative remedies; eg in their reliance upon state licensing of premises or occupations. This is a drastic technique previously found only in the legislation regulating mines, quarries, explosives, nuclear and petroleum installations. However, it is apparent that regulation-making power is now available to apply this technique to any work situation, whenever it is considered to be appropriate and effective, without the need for further specific enabling legislation. Thus the Advisory Committee on Asbestos has proposed that there should be a licensing scheme for persons working with sprayed asbestos coatings or asbestos insulation[5].

Again, the Mines and Quarries Act is unique among the earlier statutes for the strictness of its requirements which are imposed upon all levels of management and not only upon owners or employers. The powers in the new Act enable obligations of a similar nature to be imposed in the future in any work situation where they may be considered appropriate.

The width of these regulation-making powers may be contrasted with the conventional regulation-making powers given in previous health and safety legislation. In the past, ministers have normally been given powers to make regulations only to implement the provisions of individual sections of the parent Act. Thus in the Factories Act 1961, most of the general provisions contained in the 62 sections, which formed the first three parts of the Act, included regulation-making powers; eg under s2 the Minister might make regulations relating to temperature in factories. Any regulations he made by virtue of this section relating to temperature in offices would have been *ultra vires* (ie outside the powers) and therefore invalid, since the Minister's powers only extended to the control of those matters in factories. Similarly, since the Factories Act contained no provision that controlled noise in factories, regulations purporting to control noise levels might have been likewise *ultra vires*[6].

The regulation-making powers under the new Act do not appear to be confined in this way: it would appear that, if a particular regulation had the object of achieving a greater degree of occupational health and safety, the general purposes of this Act would enable it to be made validly.

In practice the regulation-making power may not be quite so wide. There may be an overlapping jurisdiction with a number of other bodies empowered to make regulations, for example the Department of the Environment under the Control of Pollution Act, the Department of Employment in relation to some of the provisions in the Employment Protection Acts and the Equal Opportunities Commission under the provisions of the Sex Discrimination Act, concerning protective employment legislation that discriminates between the sexes[7].

On the other hand the specific requirement contained in s1(2), that regulations made under the Act shall maintain or improve standards of health and safety established under previous enactments, does suggest circumstances in which regulations could be rendered invalid. Presumably if a regulation attempted to relax existing standards that regulation would be invalid, for example s14(1) of the Factories Act 1961 requires that 'every dangerous part of any machinery shall be securely fenced'. It might be that a proposed regulation replacing this strict liability by a requirement that 'everything reasonably practicable shall be done to ensure that every dangerous part of any machinery shall be securely fenced' would be held to be invalid since it would appear to relax the existing standards.

Structure of the Act

The Act is divided into five main parts and contains 10 Schedules.

By far the most important part of the Act for the purposes of this book is Part I, entitled Health and Safety in Connection with Work and Control of Dangerous Substances and Certain Emissions into the Atmosphere, and the first four Schedules which are closely related to this Part.

Part I contains the general purposes of the Act and the general duties. These duties contain the substantive rules, which together with the general purposes section, form the basis of all regulations and codes of practice made under the Act. It also contains the rules relating to the Health and Safety Commission and the Health and Safety Executive: these bodies are responsible for keeping the problem of occupational health and safety under review and for enforcing the occupational health and safety legislation. Schedule 2 of the Act contains additional provisions relating to the constitution of the Commission and Executive. (For a detailed description of these bodies see chapter 5.)

Part I of the Act also contains the rules relating to the making of regulations and codes of practice and Schedule 3 contains a list of matters which may be the subject of health and safety regulations.

This part of the Act sets out the powers of inspectors, including their powers to issue improvement and prohibition notices. Also the provisions for offences and penalties.

Part II of the Act is devoted to the Employment Medical Advisory Service and its functions.

Part III of the Act is entitled Building Regulations and Amendment of Building (Scotland) Act 1959 and bears little relation to the remainder of the Act. This part will not be considered in this book.

Part IV contains a number of miscellaneous and general provisions, largely concerned with minor amendments and repeals of earlier legislative provisions.

Transitional provisions

(Section 1(2) and Schedule 1)

The general purposes section of the Act provides that it is an objective of the Act that the enactments listed in Schedule 1, and the regulations and other instruments in force under those enactments, shall be progressively replaced by a system of regulations and approved codes of practice designed to maintain or improve the standards of health, safety and welfare established by or under those enactments (s1(2)): this work is proceeding at a slow pace.

In an interim period, possibly lasting many years, the familiar existing statutory codes are being gradually phased out and replaced by regulations and codes of practice made under the general duties or the general purposes sections of this Act. During this period it will be necessary to be constantly vigilant to keep abreast of the new regulations as they are issued. New regulations of particular significance to organizations are being reported in the relevant trade and professional journals or specialist updating systems[8]: additional advice may be obtained from the central office of the Commission and Executive or local offices of the Health and Safety Inspectorate. It is possible to place a standing order with HMSO to receive all regulations made under the legislation. Many public libraries stock Halsbury's *Statutory Instruments* in their reference sections and this may prove a helpful reference source. Whatever method of monitoring the changes is adopted, great care will have to be exercised to ensure that the information obtained is indeed up to date.

Administration and enforcement

The Act placed the administration and enforcement both of the existing provisions (that is the diminishing body of earlier legislation already discussed that remains in force until it is replaced by new regulations) and the Act itself under the jurisdiction of the Commission and the Executive. The responsibility of these agencies covers the 'relevant statutory provisions'; that is the Act, regulations made under it, the earlier statutory provisions listed[9] in Schedule 1, and regulations made under them. This legislation includes such statutes as the Factories Act 1961, the Mines and Quarries Act 1954 and the Offices Shops and Railway Premises Act 1963; all of which, since 1 January 1975, have been enforced through procedures introduced by the 1974 Act.

Importance of participation

The Robens' philosophy was that since accidents happened at the workplace much of the effort to eliminate the hazards which lead to accidents must also come from the workplace. A logical development of this philosophy is to find mechanisms which allow employer and worker participation with the government in workplace safety, while ensuring that the ultimate responsibility for selecting and operating safe systems of work remains with management. The Act reflects this idea by providing for the involvement of both employers and employees in making and monitoring safety standards. There are aspects of a consensus approach in this process of consultation, with both the strengths and weaknesses of consensus evident in the development of safety standards.

The Health and Safety Commission (HSC) is, by its constitution, representative of the social partners (s10(3)). Thus proposals emanating from the Commission are themselves consensus proposals. Further, in carrying out its functions the Commission has come to adopt a policy of going beyond the bare requirements of the Act in seeking agreement with management and organized labour before introducing new safety standards. Thus the proposals of the HSC are frequently presented as consultative documents before draft regulations or codes of practice are finalized. This implies that new standards take longer to agree, but also that they carry the weight of approval of both sides of industry.

Robens advocated that a very great contribution to the development of industrial self-regulation could be made by industry-based organizations in dealing with the particular problems of various industries. He thought these should be in the nature of joint standing committees on safety. The power for the HSC to set up advisory committees was provided for in s13(1)(d) of the Act and a number of advisory committees have been created representing employers, trade unions and independent experts, as well as members of the inspectorate, eg the Advisory Committee on Major Hazards and the Advisory Committee on Asbestos.

In addition to this the Health and Safety Executive (HSE) has recognized the need for increasing specialization within the inspectorate to match the increasingly specialized technology of industry itself and has therefore organized itself into National Industry Groups (NIGs), in place of the old system of Divisions based on geographical areas. The objectives of the NIGs have been set out as follows:

(a) to provide a centre for the collection of data about practices, precautions and standards, and to provide a stimulus to the preparation of guidance notes for industry

(b) to develop contacts with bodies representing interests in that industry: management, unions, suppliers of equipment and professional organizations
(c) to pinpoint health and safety problems in the industry, whether the problems are peculiar to that industry or general ones
(d) to develop ideas about ways of improving health and safety performance in the industry
(e) to maintain consistency of enforcement practice in relation to that industry
(f) to stimulate thinking and promote constructive and planned initiatives by the industry itself[10].

Thus high priority is given to co-operation between the HSE groups and the industries with which they are concerned for the purposes of developing safety standards.

All proposals for making regulations on a particular topic must officially emanate either directly from the Commission, or form the subject matter of consultations between the Secretary of State for Employment and the Commission (s50(1) and (2)). In this way proposals coming from either of the social partners must receive the support of HSC before they can be published. As part of the consensus approach it is HSC's custom to consider the cost to industry of implementing proposals as well as the benefit to be obtained by safer working practices.

At the drafting stage of regulations the Commission is additionally required to consult other bodies which appear to be appropriate (s50(3)(a) and (b)), these invariably include employers, employees and professional associations in the industries affected by the draft regulations. However at this stage there is apparently more emphasis upon consultation than upon arrival at a consensus since, interestingly enough, there is no longer a formal mechanism for the bodies consulted to register disapproval of the proposed regulations.

All codes of practice, whether prepared by the Commission or otherwise, require the approval of that body; so again their content must reflect management/labour consensus to a considerable extent. Before giving their seal of approval to any such draft code, the Commission must consult with the same representatives of the interests affected as in the case of regulations. (s16(2)).

At the workplace level the emphasis upon participation is again reflected in the provisions of the Act. Consultation, co-operation and communication between employers and their employees for safety purposes is made obligatory by the general duties, ss2(3), (4) and (7) (for details *see* chapter 3). The functioning of safety representatives (*see* chapter 10) is eloquent testimony of the working of the Robens philosophy of participation.

3 The general duties I: employers and employees (sections 2, 7, and 9)

The nature of general duties

One of the most important innovations in the Act was the introduction of broad general duties, imposed for the protection of persons at work from accidents or ill-health arising from their work, and for the protection of the general public from exposure to risks created by the activities of persons at work.

In concept, the general duties superficially bear a resemblance to the civil law of compensation for personal injury, having much in common with the duty of reasonable care which the common law nowadays expects each person to observe in respect of his neighbour; a duty whose content varies with the factual circumstances of each case. What the legislature appears to have done in the Act is to have reproduced the concepts of the civil law (relating to accident compensation) in the criminal law (relating to accident prevention). However, the civil law is concerned only with a person who has broken his duty *and* as a consequence caused injury to his neighbour; no steps are taken by the civil courts to prevent behaviour which has *not caused injury*, however dangerous it may have been. Under this Act such behaviour, which the judges have stigmatized in the past in compensation cases, may be punished by criminal sanctions in the cause of accident prevention, whether or not an accident has occurred.

The Act identifies a number of categories of person who could conceivably place others at risk in a work situation or whose work-related activities could place the general public at risk. Persons in these categories are placed under an obligation to ensure, so far as is reasonably practicable, that they do not place others at risk.

The significance of these general duties is their very generality.

They are in marked contrast to the specific standards which the legislature has traditionally enacted on health and safety at work, the definition and interpretation of which have been the subject of so much litigation in the past. For the time being these new general duties are imposed in addition to the earlier statutory provisions; so that, for example, at present both the Factories Act 1961 and the new general duties apply to factory premises. In such situations the Act extends and systematizes the statutory protection previously enjoyed. Thus dangerous situations which could not be regulated under the Factories Act might well be rendered unlawful by virtue of the general duties. For example, it might well be unlawful now to provide a machine with a guard which (though complying with s14 of the Factories Act) was inadequate to prevent the machine itself or the materials fed into it from flying out to injure the operator[1].

Since the general duties apply to all persons at work, they have the effect of bringing under statutory protection approximately eight million workers (such as hospital employees and school teachers) who had not previously enjoyed any legal protection and whose places of employment had hitherto been beyond the jurisdiction of any health and safety inspectorate. For the time being these workers will have few detailed or specific standards laid down for their protection but may rely upon the general duties. It may be suggested that the employer of such people might well find it expedient to comply with the minimum legislative requirements under the old safety codes whenever they are relevant. Even if these codes have not been specifically extended to cover the new areas of employment, it seems highly probable that an inspector may subconsciously, if not consciously, be led to consider the standards with which he is familiar in other employment contexts as the minimum standards with which the prudent person might be expected to comply, in order to take such steps as are reasonably practicable to provide for the safety of others.

As s1 of the Act contemplates that the old statutory codes and regulations made under them will be reviewed and replaced by fresh regulations, it is clear that in due course the general duties will emerge as the only statutory duties relating to health and safety at work. All particular standards will then be found in the regulations, made either to implement the general duties themselves or, independently of the general duties, to implement the general purposes of the Act.

It should be remembered that, while the general duties may in time be used as the basis of detailed regulations dealing with specific situations, these regulations are unlikely to be exhaustive. The general duties are being enforced, and are likely to remain enforceable, in their own right[2], either by the issue of notices requiring the person upon whom the notice is served to take action to comply with his statutory duty, or by means of a prosecution.

General duties and private rights

The breach of one of the general duties does not in itself give rise to a right of action in civil proceedings. Accordingly an injured employee will not be able to bring an action for breach of statutory duty against his employer on the ground that the latter is in breach of his duty under s2 of this Act: or against the manufacturer of an unsafe machine on the ground that he is in breach of his duties under s6. Civil disputes of this nature will have to be resolved according to the earlier law, either as breaches of statutory duty under the earlier statutory provisions or, in some cases, by actions for breach of contract. For the same reasons the employer will not be able to rely on s7 where an employee causes him loss through negligence[3].

The existence of the duties under this Act may affect the creation and enforcement of contractual rights. The employer in particular may insist upon rigorous contractual terms from manufacturers, suppliers and subcontractors. He may also consider that his duties to instruct and supervise employees make it desirable that, for his own protection, he enforces rigorously against his employees his express or implied contractual rights affecting discipline particularly in the context of compliance with safety rules. There have, for example, been a number of cases in which it has been deemed to be 'fair' to dismiss an employee for misconduct related to safety[4]. Employees for their part may be more ready to take individual[5] or collective action in support of contractual rights in relation to health and safety at the workplace.

'Reasonably practicable'

(Sections 2–6 and 40)

The scheme of the general duties is that the person upon whom such a duty is imposed is required to do all that is 'reasonably practicable' to ensure the safety of those protected by the duty: thus for the most part these duties impose something less than a strict liability[6].

The civil courts have given much valuable guidance on the meaning of the expressions 'practicable' and 'reasonably practicable'. The former is generally taken to mean something less than 'physically possible': that is, possible in the light of current knowledge and invention[7]. Thus there is no fixed standard. Situations and practices which were formerly acceptable may become unlawful as advances in knowledge and technology make new standards appropriate. In considering whether it was 'reasonably practicable' to avoid a risk the courts may well have regard to the cost of the preventative action when weighed in the balance against the probability of personal injury occurring and the severity of the injury likely to occur[8]. The courts will, however, probably apply the test objectively and are unlikely to have much sympathy with such special considerations as the financial resources available to the particular defendant or his *bona fide* ignorance of the standards in question. In fact this is precisely the way in which the matter has been approached by the industrial tribunals in considering appeals against enforcement notices (*see* chapter 7).

These qualified duties may be contrasted with the 'strict' duties sometimes imposed under the earlier safety codes (such as the duty imposed on the occupier to fence dangerous parts of machinery under the Factories Act 1961) which remain in force for the time being. It is highly likely that new regulations imposing strict liability for the safeguarding of machinery will continue to be issued under the Act, since s1 calls for regulations that maintain and improve safety standards. The regulation-making powers (discussed in chapter 6) are wide enough to achieve this purpose.

It does not follow that the general duties are less than onerous. Section 40 provides that in any proceedings for an offence under the relevant statutory provisions consisting of a failure to comply with a duty, or requirement to do something so far as is practicable, or so far as is reasonably practicable, or to use the best practicable means to do something (all expressions to be found either in the Act or in the earlier statutory provisions) it shall be for the accused to prove (as the case may be) that it was not practicable, or not reasonably practicable, to do more than was in fact done to satisfy the duty or requirement, or that there was no better practicable means than was in fact used to satisfy the duty or requirement.

In other words, once it is proved that a person has failed to prevent the existence of an unsafe situation, the burden of proof will shift to that person to show that he could not have taken suitable steps to eliminate that danger. Thus once it is established, for example, that the floor of a factory was obstructed contrary to the employer's duty under s2(2)(d) of the Act (and for the time being, the occupier's duty under s28 of the Factories Act 1961) it is for the employer or the occupier, as the case may be, to establish that it was not reasonably practicable for the place to have been kept free from obstruction. If he cannot discharge this burden of proof[9] he will be liable. Section 40 of the new Act therefore gives statutory approval (for the purposes of all criminal proceedings, under the Act and the relevant statutory provisions) to a rule previously established in civil actions arising under earlier safety legislation[10]. This rule applies both to proceedings for safety offences before justices and in trials upon indictment in the Crown Courts[11].

It should be noted that this burden of proof applies only in proceedings in relation to an offence; disputes as to the steps which are required to be taken to comply with the duties imposed by the Act and regulations may be contested before industrial tribunals when the validity of improvement and prohibition notices are being contested. In the somewhat less formal atmosphere of the industrial tribunals, these rules relating to burden of proof may well not apply and formality may not be so strictly observed[12].

The general duties are set out in sections 2–8 of the Act: they impose separate requirements upon employers, self-employed persons, controllers of premises to which the Act applies, designers, manufacturers, importers and suppliers of articles and substances for use at work, and employees while at work. In view of the importance of each of these provisions they are dealt with individually in separate sections.

Duties of employers

(Section 2 and s9)
Section 2(1) of the Act imposes a comprehensive duty upon every

employer to ensure, so far as is reasonably practicable, the health, safety and welfare of all his employees. A similar duty in respect of persons not in his employment, who might be affected by his activities, is imposed by s3(1). Section 2(2) goes on to give examples (which are not necessarily exhaustive) of particular aspects of the duty of the employer to his employees. An analysis of these examples indicates clearly that a great deal of responsibility rests upon management at all levels to ensure, as appropriate, that the necessary organization has been set up and functions satisfactorily, to comply with these duties at all times. In particular, management must ensure that plant and equipment are safe, and that safe systems for the prevention of accidents and for the protection of health are in operation.

(a) Plant and equipment

Section 2(2)(a) requires the provision and maintenance of plant and systems of work that are, so far as is reasonably practicable, safe and without risk to health. In the light of this requirement, it may well be appropriate for management to keep under review not only the equipment itself (its age, reliability, appropriateness for the task in the light of advancing technology) but also its layout and maintenance.

Managers should ensure that their operatives are sufficiently trained to detect faults in the machinery they use and instructed in how to rectify these faults, either personally, or by obtaining the assistance of others. Also managers should satisfy themselves that they have provided an effective system of maintenance, either by employing trained maintenance engineers or by arranging suitable subcontracting. The acid test would seem to be whether faults are likely to be detected and remedied as soon as possible, even if this may involve an interruption in production in the meantime.

(b) Handling, storage and transport

Section 2(2)(b) requires an employer to have suitable arrangements to ensure the safe use, handling, storage and transport of articles and substances. The Robens terms of reference specifically excluded matters relating to transport workers while directly engaged on transport operations. Although Parliament was not obliged to confine the legislation to the Robens terms of reference, it seems unlikely in practice that this section would be used as the basis of regulations relating to the conduct of drivers on a highway, since this falls more appropriately under the Road Traffic Acts. But it does seem desirable that, in the light of this requirement, an employer should have particular regard to the instruction and training of his employees in safe systems of handling goods, especially those in force on his premises. An employer might well keep under review, for example, his system for loading and use of fork-lift trucks, whether gangways and exits are kept clear of goods, whether adequate lifting equipment is provided for handling heavy materials and whether goods are safely stacked for storage. It should also be remembered that the Road Traffic Acts do not apply to the movement of vehicles on private premises; thus the Act might be invoked to deal with dangerous practices or accidents arising out of the movement of vehicles on premises where persons are at work[13]. Apart from these general considerations as to the conduct of workers on both private premises and the public roads, specific regulations have been made governing the handling, storage and transport of solid and liquid hazardous substances[14].

(c) Systems of work, including information, training and supervision

The duty under s2(2)(a) to provide a safe system of work is also of importance to the personnel manager, particularly when considered in the light of s2(2)(c), which requires the employer to provide such information, instruction, training and supervision as is necessary to ensure, so far as is reasonably practicable, the health and safety at work of his employees.

These requirements are of fundamental importance and are wide in scope. They go to the heart of the Act, because they require the provision and maintenance of a system which ensures people are equipped as individuals to operate safely, both for their own protection and for the protection of others. It should be noted that the Act does not limit itself to requiring the provision to individual employees of information, instruction and training for their personal safety. The provision in s2(2)(c) is much wider and clearly envisages a complete organizational structure for disseminating information and providing instruction and training throughout the whole management structure, in order to ensure that each person is equipped to provide not only for his own safety but also for the safety of others over whom he exercises managerial responsibility or with whom he works. This provision is wide enough to require an employer to provide, in some situations, information about safety precautions to persons who are not even in his employment[15]. While it is clear that it is within the spirit of the Act that employees should be informed of the hazards to which they are exposed, it is equally clear that mere information, without the provision of a managerial system for the operation of the safest systems possible to cope with the particular hazards, would not constitute a discharge of his duties by the employer. It is important to note that these requirements are quite independent of the employer's duty to provide certain information to safety representatives under the regulations. The rights and functions of safety representatives are discussed in chapter 10.

(d) Identification of hazards

The positive identification of risks is fundamental to the provision and maintenance of a safe system of work. There is no doubt that this requirement imposes an onerous burden upon management; so that there can be no question that (to comply with s2(2)(c)) an organization would be deemed to have knowledge of, and therefore be under a duty to inform employees of risks that were common knowledge in the trade in which the employer is operating. In addition to this, however, the employer might well be required to carry out tests of his own if, for example, he is introducing a method of production or employing a material that is not commonly used in his trade.

The discharge of this aspect of the employer's duty would seem to call for close cooperation between specialists and line management. In some cases difficult management decisions may have to be taken to steer the narrow path between, on the one hand, spreading unnecessary alarm by the dissemination of incompletely verified information and, on the other, placing persons at unnecessary risk by withholding relevant information. For example, research might suggest a possibility that certain processes could be a cause of skin cancer. In such a case the employer would be required to do all that is reasonably practicable for him to do, to establish the truth of this proposition, but he might well be in a dilemma as to the moment at which suspicion becomes sufficiently near to certainty to warrant disclosure[16]. This problem is likely to be particularly acute in, but by no means confined to, establishments in which employee safety representatives are functioning.

It should be noted that disclosure and consultation are of the spirit of the Act, and the duty of management to inform its employees of risks to which they are exposed is closely related to the duty of the inspector under s28(8) (discussed in chapter 9) to inform workers, or their representatives, about matters affecting their health and safety. This latter duty relates to any factual information about the premises in which the employees are employed or anything which is in them or is being done therein. It will generally be discharged by communication to safety representatives, but it arises in all workplaces whether or not safety representatives have been appointed.

(e) Periodical medical examinations

A careful decision may also have to be made as to whether the duty imposed by s2(2)(c) to provide information and supervision may be interpreted merely as a duty to supply health information to employees as a group, or whether it requires the monitoring of each individual employee's health, for example, by periodical medical examinations. Case law has already suggested that, where compensation issues are involved, the courts are prepared to impose upon an employer a duty either to arrange for medical examinations for individual employees who are particularly exposed to risk, or at least to take steps to ensure that employees can monitor their own health and recognize when to go to a doctor[17].

(f) Instruction and training

The express inclusion of instruction and training within s2(2)(c) should certainly give the employer cause to review the adequacy of the instruction and training given to his employees at all levels to ensure their own safety and the safety of other employees. Clearly the discharge of this aspect of the employer's duty requires, initially, the instruction of managers in the necessary techniques and systems for the recognition and control of risks, for the safety of the workforce. Subsequently all operatives must receive the skills training necessary for the safe performance of their tasks. It should also be stressed that 'instruction' in law implies discipline and enforcement; the employer who had on paper an adequate system which he failed to enforce in practice would hardly be likely to be deemed to have complied with the requirements of this section. It therefore falls to management to exercise constant vigilance, not only to see that employees are properly trained to operate safely but that in practice they do observe safe systems for the sake of themselves and others. The discharge of this aspect of the employer's duty means that line management must not only be aware of the responsibility to upbraid those who disregard safe practices but be confident that they have the authority to do so and that disciplinary measures, initiated at the level of first line management will, if warranted, be supported by superior management.

In many instances this duty to instruct and train at all levels cannot be discharged from within the resources of an organization. Here, compliance with the law will necessitate the use of the expertise of outside training organizations.

(g) Safe premises

Section 2(2)(d) requires the employer to do all that is reasonably practicable regarding any place under his control, to maintain it in a safe condition and to maintain safe access and exit from it. This duty concerning the premises in which the employee is required to work may be compared with the similar, but more specific, duties under the Factories Act 1961, s28 and s29. It should also be compared with s4 of this Act which places a duty upon the controller of the premises in respect of visitors to his premises who are not his employees.

(h) Safe working environment

Section 2(2)(e) requires from the employer the provision and maintenance of a working environment for his employees that is safe as far as is reasonably practicable. The wording of this paragraph would appear to require more than regard merely to the physical environment of the worker (which in any case is largely covered by s2(2)(d)). Here, management would be well advised to consider the emotional satisfaction which it is possible for the employee to obtain from his employment and to have particular regard to the problems of mental stress and the practical steps which can be taken to alleviate it.

(i) Statement of safety policy

It is the employer's duty not only to provide a safe system of work for his employees, but also to make clear to them his commitment to their health and safety at work and the means by which he proposes to safeguard them.

Section 2(3) requires the employer (unless exempt by regulations)[18] to prepare a written statement of his general policy on the health and safety at work of his employees. He must also set out the organization and arrangements for the time being in force for carrying out that policy. He must bring the statement and any revision of it to the notice of all his employees.

While some employers may initially have tended to regard these requirements as a piece of bureaucracy, those who have conscientiously attempted to comply with the Act have found compliance no mean task, particularly if their organizations are large and complex. One cannot help thinking that because this provision requires that each employer give positive thought to his own organization it may be quite crucial to meeting the objectives of the Act.

The subsection has two requirements: first that the employer state his policy and, secondly, that he set out his organization and arrangements for carrying out his policy. The first requirement of the section is not difficult to comply with, and can in fact be done merely by adopting the words of s2(1) of the Act itself in which the general duty of the employer to his employees is set out. Many employers however may think it desirable to do more than the section strictly requires and therefore state their policy regarding other persons at work and the general public, as well as that related to their own employees. In practice, the more specific the information relating to the actual hazards likely to be encountered at the employer's workplace, the more valuable the policy will be. In some instances detailed hazards information may be supplied in supplementary documents, eg safety manuals[19], referred to in the general policy statement. Many employers have also taken the opportunity of stating what they expect of their employees as their contribution to safety at the workplace. This again goes beyond the requirements of s2(3) and, although valuable in directing employees' attention to their responsibilities, cannot of itself legally impose either statutory or contractual obligations upon an employee. It must also be stressed that while an employer may discharge his obligations under this subsection by making provision for a safe system of work in his safety policy, he cannot use the policy as a means of delegating to managers or more junior employees his own legal responsibilities under other provisions[20].

The real sting in s2(3) is the requirement that the employer state his organizational structure for carrying out his policy. Many employers have found this an extremely difficult exercise. Moreover, it is one upon which they can expect little practical guidance since individual arrangements must necessarily depend upon the individual employer's management structure. It is possible that some employers have erred by attempting to create a new management hierarchy for dealing with the execution of their safety policy, rather than considering how setting out a safe system of working is related to their existing management structure. It is doubtful whether a safety policy could be considered adequate if it did not state the responsibilities attaching to the various posts within the management hierarchy.

In many instances consideration of these responsibilities in relation to the existing management structure has served to bring to light faults in the structure which were not necessarily confined to issues of safety. Writing a safety policy may therefore serve as a review exercise of the definition of responsibilities and the identification of the links in the management chain generally, and may well reveal weaknesses in the channels of communication important to good management. It would seem unlikely that a firm could achieve a good system of safety within a management structure which was generally bad.

The safety policy statement when made will have to be kept constantly under review and may need revision from time to time to bring it into line with changing circumstances. Several years have now elapsed since the requirement for safety policies was first introduced and management practices may have changed; many organizations have made significant changes in their management structure and even in many firms, safety policies (initially distributed to all employees) are now forgotten. So for these and many other reasons any firm which has not reviewed its original safety policy would be well advised to do so[21].

It is the employer's duty to bring the statement (and revisions of it) to the notice of all his employees. The method adopted is left open to the employer, but it is suggested that the employer who does not provide each employee with a personal copy of the statement will have to establish that the steps taken to bring it to the notice of all employees are effective ones. It may be sufficient to provide employees with a note informing them where the statement may be read; it is doubtful if anything less than this would be effective communication on the part of the employer. Merely to place the statement on a notice board is of doubtful value, particularly if (which is not desirable anyway!) the statement is long and involved. The statement should also be discussed with the representatives of recognized trade unions: these representatives may well wish to be included in any revision of the policy.

In the case of a group of companies, the parent company will probably state the policy and name the officers of subsidiary companies who are responsible for making organizational arrangements for each part of the group. Something similar may be desirable between head office and the various sites in an organization which operates in a number of different geographical areas.

Many companies have deemed it expedient to put out a policy statement which has been signed by a member of senior management. This may well be a useful way of indicating the importance which the employer attaches to his safety policy but it is not a requirement of the Act.

(j) Safety policies and directors' reports

It should not be forgotten that the Act also includes provisions for reporting corporate safety arrangements to a larger audience.

Section 79 amended the Companies Act 1967 to empower the making of regulations requiring the inclusion within directors' annual reports of the prescribed information as to companies' safety arrangements.

This information will relate to the arrangements in force for the year under review for securing the health, safety and welfare at work of the employees of the company and its subsidiaries and for protecting other persons against risks to health and safety arising from the activities at work of those employees.

The regulations will be applicable to such classes of companies as may be prescribed: a likely classification would be by size or type of activity and in any event different reporting requirements may be enforced for different classes of companies[22].

(k) Consultation

The first page of the first chapter of the Robens Report contains the statement '... the most important single reason for accidents at work is apathy'. Robens placed worker involvement high in the list of recommendations for the improvement of the situation, saying of work people '... they must be able to participate fully in the making and monitoring of arrangements for safety and health at their place of work'.

Under s2(4) regulations may require in prescribed cases the appointment by recognized trade unions of safety representatives from among employees of the employer to represent employees in consultation with the employer.

A parallel subsection (originally s2(5)) made provision for the election by employees in prescribed cases of safety representatives from among the employees to represent the employees in consultation with the employer: however, this provision was repealed by the Employment Protection Act 1975.

Section 2(6) imposes a duty on the employer to consult the appointed representatives with a view to making and maintaining the arrangements which will enable him and his employees to cooperate effectively in promoting and developing measures to ensure the health and safety at work of the employees, and such other functions as may be prescribed. These arrangements may include the setting up of safety committees.

Regulations were made under s2(4) in 1977. The implications of the implementation of these regulations is so great that in chapter 10 the system which they have introduced is explained in detail. A full discussion of the rights and responsibilities of workers under the Act will be found in chapter 9.

(l) Duty not to charge employees

Section 9 provides that no employer shall levy or permit to be levied on any of his employees a charge in respect of anything done or provided in pursuance of any specific requirement of the relevant statutory provisions. The section is in the same spirit as s136 of the Factories Act, which prohibited deductions from wages in respect of anything provided by the occupier of the factory in pursuance of the specific requirements of that Act. Section 9 of this Act could, however, be interpreted much more widely and thus some doubt is created as to the occasions upon which an employer can lawfully charge for the provision of protective clothing. There can be little doubt that if the employer provides, say, eye protection in compliance with regulations which specifically require its provision he cannot charge for this equipment, whether the regulations concerned were made under the Factories Act or s2(1) of this Act. The 'grey area' is where, for example, the employer provides ear muffs because these seem to him, in his discretion, to be the most satisfactory way of protecting the hearing of his employees – which he is bound to try to do to discharge his general duty under s2(1) to ensure so far as is reasonably practicable the health, safety and welfare of his employees. Since the employer provides the ear muffs in discharge of his general duty, can he charge for them? A reasonable interpretation of s9 would appear to suggest that he could, since the ear muffs, unlike the eye protection in the previous example, are not supplied in pursuance of a specific statutory requirement but only because, in his discretion, the employer elects to supply them as one appropriate way of discharging his general duty. In view of the element of doubt[23], employers may well deem it expedient to review their policy if they are at present charging for protective clothing particularly if they wish to enforce the use of this clothing in discharge of their general statutory duties under s2.

The duties of employees

(Section 7)
Section 7 imposes two duties upon the employee while at work; the duty to take reasonable care for the health and safety of himself and of other persons who may be affected by his acts or omissions at work, and secondly the duty to cooperate with other persons to enable those other persons to carry out the statutory duties imposed upon them in respect of health and safety at work.

The imposition upon an employee of these wide duties, to take reasonable care and to cooperate with others, are among the more radical innovations of the Act[24]. The comparable provision of the Factories Act 1961 in s143 was much narrower in its terms, being aimed only at wilful action on the part of the employee. Section 143 was ineffective as it was rarely invoked by factory inspectors, and only 21 employed persons were prosecuted for its breach in 1973. The limited use of s143 may have been due as much to a reluctance to prosecute employees, (particularly in cases where they had already suffered the 'punishment' of personal injury as a result of their misdoings), as to the restrictive terms of the statutory provision itself.

At manual worker level, the first part of s7 could be invoked against an employee who removed the guard from his machine, or a driver who used his fork-lift truck without proper regard for the safety of others. It might also be used where horseplay creates danger[25].

The duty to cooperate with the employer or others in discharge of their statutory obligations is, moreover, very wide and certainly not confined to manual workers. Managers below board level are all within the definition of employed persons and the subsection could be invoked against managers who failed to exercise proper diligence in carrying out their managerial functions by failing, for example, to institute and maintain safe systems of work.

It would appear that the higher up the management scale one goes the wider the span of responsibility and the greater the possibility of being held to account under these provisions[26]. It may well be that senior employees are particularly vulnerable in respect of the second limb of the section which makes it an offence to fail to cooperate with others in the performance of their duties.

The employer himself cannot use s7 as a means of disciplining recalcitrant employees, since under the Act only an inspector may institute a prosecution in England and Wales. Therefore the importance of this section must depend upon the policy adopted by the inspectorate – as indeed must be the case in respect of all the general duties.

Offences committed by employees

There may well be occasions when s7 will have to be considered in association with s36 which provides that, where the commission of an offence is due to the act or default of some other person than the one upon whom the legislation has imposed the duty, that other person shall be guilty of the offence.

Take for example the situation in which one worker is injured because another worker has created a dangerous situation, perhaps having taken a fence from a machine: the state of the machine could create *prima facie* liability, on the employer for breach of his duty under s2. The wrong doing worker could be liable under s36 for creating the situation which *prima facie* involved his employer in breach of s2. Alternatively that employee might also be liable for breach of the duty imposed on him directly under the first part of s7.

The Chairman of the Health and Safety Commission has indicated that as the Crown as employer is immune from prosecution under the Act, in his view it would not be equitable to prosecute Crown employees under ss7 or 36. This is rather surprising since s48(2) specifically provides that the Act shall be enforceable against Crown employees despite the Crown's 'personal' immunity from liability. (For enforcement of the Act in Crown premises, see chapter 7).

4 The general duties II: other persons (sections 3, 4, 5, 6 and 8)

Shared duties of employers and self-employed persons

(Section 3)
Section 3 places a duty upon every employer to conduct his undertaking in such a way as to ensure, so far as is reasonably practicable, that persons not in his employment who may be affected by the undertaking are not exposed to risks to their health and safety.

It also requires every self-employed person to conduct his undertaking in such a way as to ensure, so far as is reasonably practicable, that he and other persons, not being his employees, who may be affected by the business, are not thereby exposed to risks to their health and safety.

The section appears to have a threefold object. First, it imposes a duty upon employers and contractors to safeguard workers not in their employment: secondly, its aim is apparently to create safe systems of work when two or more organizations or individuals are each conducting part of a joint operation, or are conducting separate operations in close proximity to each other; and thirdly, to protect the public.

These provisions taken together represent an attempt to rectify some of the more glaring deficiencies of the older law such as the Building (Safety, Health and Welfare) Regulations 1948, made under the Factories Acts. Under such regulations the duties of an employer did not necessarily extend to the provision of safe working conditions for self-employed persons who were sub-contractors using the employer's equipment; equally they imposed no obligations upon the self-employed person who employed no labour. Thus the latter was left at liberty to disregard totally both his own safety and that of others, who might well suffer injury without redress as a result of relying upon unsafe equipment, brought on to a site, or erected thereon, by the self-employed person. Thus such legislation not only left individual workers and the public without protection, but it also made no attempt to provide for a safe system of work.

An effect of s3 is clearly to impose liabilities upon subcontractors who do not adopt a safe system for the coordination of their respective duties. Mutual responsibilities in this type of situation were recognized for compensation purposes before the Act. Thus a head contractor, a subcontractor and the injured man's employer (another subcontractor) shared the liability to compensate him[1]. Since the Act a head contractor incurred criminal liability under the section for failing to inform subcontractors of the dangers of his workplace and the measures to contain them[2]. Similarly he would be liable if he engaged two subcontractors and failed to discuss with them the coordination of their respective activities. Nevertheless the liability of the head contractor did not exonerate the subcontractor from liability in that case: indeed the subcontractor could have been held liable either for failing to identify and adopt appropriate systems of work himself under ss2 and 3 or for failure to follow the systems of which he was notified by the head contractor (s36).

For the purposes of both the criminal and the civil law a firm which, for example, engages a contractor to paint its building will be well advised not only to discuss with the contractor the provision of a safe system of work on the firm's premises, but to make compliance with the agreed system an express term of the contract between them. If this is done the firm (and indeed the contractor) will be in a strong position either to enforce the agreement or to repudiate the contract if the other party engages in unsafe practices contrary to the agreement. Thus the Act, even if it cannot be said specifically to require a firm's contract department to apply entirely new criteria when making contracts with subcontracting organizations, now gives extra significance to practices which ought to have been firmly established for many years. The sanction in the past was the possibility of liability to pay compensation to injured parties, and could have been effectively provided for by contractual provisions concerning insurance – the sanction now may be criminal and therefore cannot be contracted away. Where hazards are encountered, the 'innocent' parties must now enforce the provisions of the contract to escape criminal liabilities. Moreover effective enforcement may necessitate prompt action against the individual employees of subcontractors.

The section is also intended to protect the general public (in accordance with s1(1)(b)) and, incidentally, gives the inspectorate enforcement powers in respect of public safety which they lacked under previous legislation. They could by virtue of this section require an employer to remedy a situation which creates a public hazard. For example, require him to reduce the noise level emanating from his factory to the street outside, or make safe a coal tip which endangers the public, or re-site a crane so that there is no danger of it collapsing upon the public on the highway outside the boundary of the building site[3].

The question of the protection of the public has become a major issue since the Act and so is considered further in chapter 12.

Section 3(3) relates to the provision of information. In such cases as may be prescribed, it shall be the duty of every employer and every self-employed person, in the prescribed circumstances and in the prescribed manner, to give to persons (not being his employees) the prescribed information about such aspects of the way in which he conducts his undertaking as might affect their health and safety. Up till now no regulations have been made under this subsection so its value remains a matter for speculation. However it should be noted that there is nothing in this subsection to restrict it to the protection of persons at work; when regulations are made to implement this provision they may well relate to the provision of information to protect the public against risks created by industrial activity. It may well be that under this subsection, information may be required to be published about neighbourhood risks created by the emission of chemical and other hazards from industrial premises into the atmosphere, to the risk of those who live nearby. In the meantime, the Court of Appeal has suggested that regulations made under s3(3) are likely to deal with very specific situations and the lack of them does not detract from the general duty which exists under s3(1) for an employer to notify visitors to his premises of hazards which they may find there[4].

Duties of controllers of premises to visitors

(Section 4)
Section 4 places duties upon controllers of premises for the benefit of those who are not their employees, but who use non-domestic premises as a place of work, or as a place where they may use plant or substances made available to them for their use.

The duty imposed upon the controller of the premises by s4 is to take such measures, as it is reasonable for a person in his position to take, to ensure so far as is reasonably practicable that the premises, all means of access thereto or egress therefrom which are available for use by persons using the premises and any plant or substances on the premises, or provided for use there, are safe and without risk to health (s4(2)).

The protection extends to both visiting workers who use the premises as a place of work and to visitors who are not workers, but who use plant or substances provided for their use on the premises, for example, visitors to a laundrette or children's playground.

A person who has control of premises 'to any extent' must comply with this section: it follows that different persons may conceivably have control of the same premises under this section for different purposes and possibly more than one person could have control at one time for the same purpose. Thus in a leading civil case a brewery and a publican shared 'control' of a public house but their legal responsibilities for the condition of the premises were different[5]. Under s4(4) it is made clear that the section is primarily directed towards the controller who occupies for the purposes of carrying on a trade, business or other undertaking, but a landlord with repairing obligations is treated as 'controller' of the areas that he is bound to repair (s4(3)). However domestic householders are expressly excluded by s4(1)(b).

The controller of the premises upon whom these obligations are laid may himself employ workers on the same premises and will then have to comply with s2 in respect of those premises for the protection of his own employees. This latter obligation, however, in no way affects his responsibilities under s4, though it may well be that in many instances by complying with the greater requirements of s2 he will have also complied incidentally with s4. The greatest significance of s4 is therefore in respect of premises at which the occupier does not normally employ persons: for example a church hall let for rehearsals to a group of professional actors, or premises where the occupier provides equipment solely for use by persons other than his own employees: for example, a cradle kept for the exclusive use of window cleaners. It is clear that in such circumstances the section imposes a duty that unlike the equivalent civil liability, may not be modified by contract: and that this remains true whether or not the public make payment to enter the controller's premises.

The scope of s4 is not entirely clear, though it is arguably more narrow than the scope of the occupier's civil law duty to exercise reasonable care for the safety of all lawful visitors to his premises. The protection of s4 is limited to visitors who are attending the premises whether to work there or to use plant or substances provided for their use there. It would appear therefore, to suggest some examples, that the section would not apply to patients attending a dentist's surgery for treatment, but might possibly apply in respect of in-patients in a hospital; it would apply to customers at a self-service laundrette, or customers for whom wire baskets are available in a supermarket, or those using crockery in a café, but it seems extremely doubtful if it would apply to a conventional retail shop with counter service.

An effect of s4 may well be to create a shared responsibility for safety between the employer of a visiting worker and the controller of premises at which he is working. We may see re-enacted under the Act a demarcation of the responsibility of the employer from that of the controller of premises for the visiting workman's safety similar to the developments which have taken place in the civil law in compensation cases[6]. It would appear that while the Act places a duty upon controllers of premises in respect of the premises themselves, and the equipment therein, this duty must be read in conjunction with the employer's duty under s2. Thus the employer who sends his employees to work on premises which he does not occupy (including in this instance domestic premises) will not have done all that is reasonably practicable for the safety of his employees, if he has failed to instruct them in the identification of, and methods of coping with, hazards incidental to their employment in these premises.

This division of responsibilities is of particular importance where warehouse staff, window cleaners, visiting and domestic staff, such as 'home helps', are concerned as workers of these descriptions are rarely the employees of the controller of the premises at which they work.

One of the most difficult questions in practice, under ss3 and 4 is the extent to which the occupier of premises should attempt to control the behaviour of visiting workers who are not his employees. Clearly the occupier will be well advised to ensure that the visitors keep to the contractual provisions concerning the performance of the work, and obey any particular safety procedures that he has requested but it is suggested that it is unwise to attempt to exercise control over the way in which the visitor performs the skills pertaining exclusively to his trade. Thus the occupier might reasonably object to a ladder being placed over a pathway, but whether he would be advised to object to a worker mounting a ladder which is neither footed nor lashed, is questionable. It is probably a matter of fact depending on the severity of the incident and the number of occasions upon which the visiting worker shows apparent disregard for his own safety, and the likely effect of unsafe working practices upon the morale of other workers, whether the occupier should intervene or not[7].

Duties in respect of harmful emissions

(Section 5)
Section 5 places a duty upon the controller of premises of classes prescribed by regulations to use the best practicable means for preventing the emission into the atmosphere, from the premises, of noxious or offensive substances and for rendering harmless and inoffensive such substances as may be emitted. Controllers of premises for the purpose of this section are persons who are in control of the carrying on of an activity on the premises; the difficulties discussed concerning the control of premises for the purposes of the last sections are unlikely to arise under this section.

The section permits the creation of duties for the prevention of neighbourhood risks and is for the protection of the general public rather than employees. It is not a duty of general application, and it has not yet been implemented by regulations. Its use is likely to be supplemental to the powers given to local authorities under the Control of Pollution Act 1974. Provisions may be made to replace the Alkali etc Works Regulation Act 1906 which is one of the earlier statutory provisions and therefore, by virtue of the general purposes of s1 of this Act, subject to repeal.

The borderline between emissions subject to the Act and those subject to the Control of Pollution Act 1974 is not clear. A reasonable possibility is that the Act will ultimately seek to control those emissions accepted as a possible health hazard to employees or the public, while the Control of Pollution Act is reserved for the regulation of emissions that are regarded as damaging to the environment or to public amenity. The duty imposed under this section is to utilize the 'best practicable' means. This expression is not found elsewhere in occupational safety legislation but is found in environmental protection legislation such as the Alkali Acts and has been the subject of judicial interpretation in this context.

Duties of manufacturers, suppliers and others (s6)

(a) Manufacturers and suppliers

Section 6 of the Act places wide and novel duties upon persons who design, manufacture, import or supply articles or substances for use at work.

Articles for use at work is a broad term which includes, for the purposes of the Act, both plant and its component parts; substances for use at work is even wider, meaning any natural or artificial substance, solid, liquid, or in the form of gas or vapour.

The basic duty imposed under s6 in respect of articles and substances supplied for use at work is to ensure that they are designed, constructed or manufactured, so as to be as safe as is reasonably practicable when properly used; to carry out or arrange for the carrying out of such testing and examination as may be necessary for the performance of this duty, and to ensure that adequate information is made available about the use for which it has been produced and tested and the conditions under which it should be used.

A further duty is imposed on the person who undertakes the design or manufacture of an article, (s6(2)) or the manufacture of a substance (s6(5)) to carry out, or arrange for the carrying out, of any necessary research with a view to discovering and, so far as is reasonably practicable, eliminating any risks to health or safety to which the use of the article or substance may give rise.

Under s6(3), it is the duty of any person who erects or installs any article (but not a substance) for use at work in any premises where that article is to be used by persons at work, to ensure so far as is reasonably practicable that nothing about the way in which it is erected or installed makes it unsafe or a risk to health when properly used.

In the main, the duties relating to 'articles' and 'substances' are the same and can be discussed together, although the implications for management will not be the same in each case. One most important difference relates to the information that must be supplied to the user, as the type of testing and examination will clearly differ for engineering plant and, for example, chemical substances which are potentially carcenogenic.

The Robens Committee placed great emphasis on the need for research before new chemicals or other substances were marketed[8], hence the sweeping nature of the research requirements of the section, which could be seen to be the equivalent of the duty imposed on manufacturers in earlier civil cases[9]. It is significant that in the period since the Act was brought into force, substantial measures have been introduced to implement the section, which is emerging as a focal point for issues relating to consumer, public and occupational safety. The growing awareness of chemical hazards has made some of the matters, to which s6 relates, of international interest and some of the regulations made under this section have stemmed from the need for national legislation to comply with EEC directives (eg labelling of dangerous substances, transport of dangerous substances). Similarly the proposal for notification to HSE of information concerning hazardous substances, prior to their being used by industry, is related to EEC thinking on this subject.

(b) Designers of articles

The same basic duty under s6 is also imposed upon the designer of an article. It is noteworthy that no equivalent duty is imposed upon the person who 'designs' a substance. But the manufacturer of a substance is bound by the stringent duty to undertake any necessary research, or to make himself familiar with the results of research as prescribed in subsections 5 and 6[10].

One effect of these provisions may be to ensure that the purchaser who designs a product to his own requirements takes upon himself the responsibilities of the designer under the subsection and may to that extent relieve from responsibility the manufacturer who produces the commodity faithfully to the specification.

The manufacturer's position in these circumstances would appear to be further protected by s6(6) which provides that a person need not repeat testing examination or research which has been carried out otherwise than by him or at his request, in so far as it is reasonable for him to rely on the results thereof. This subsection would appear to entitle the manufacturer to rely on the accuracy of the work of the designer in cases where the functions of designing and manufacturing are separated.

(c) Manufacture or design to customers' own specifications

Under s6(8) it is provided that where a person designs, manufactures, imports or supplies an article (but not a substance) for or to another on the basis of a written undertaking by that other, to take specified steps sufficient to ensure, so far as is reasonably practicable, that the article will be safe and without risk to health when properly used, the undertaking shall have the effect of relieving the manufacturer, designer etc from his duty under s6(1)(a) to ensure that the article is so designed and constructed as to be safe and without risk to health when properly used.

This subsection might be used to advantage by the manufacturer who retains doubts as to whether the purchaser who has drawn up his own specification has had a sufficient regard to the safety of the persons by whom the article is to be used at work. It should be noted that the effect of the subsection is limited because it only applies where a written undertaking has been obtained and only in respect of specified steps. It could not therefore be used as a general disclaimer of liability. In any event it is questionable whether the provisions of this section could effectively be covered by a clause in a standard form contract imposed by the vendor upon the purchaser, apart from the more general point that there is often considerable doubt whether such clauses incorporated in such places as the back of an order form, constitute valid terms of a contract even if they relate to that ever diminishing range of liabilities in respect of which the law permits the parties to exempt themselves.

Quite apart from s6(8), difficult problems of division of responsibilities could arise where, for example, an employer orders a machine or substance from a manufacturer's stock, stating that it is required for a particular purpose. It would appear that in such a case the supplier would be wise to ensure that he is discharging his statutory duty by informing the purchaser of what, in his view, is the proper use of the commodity ordered, and expressly exonerating himself from responsibility for the goods if used for the purposes required by the purchaser, at least by drawing the purchaser's attention to his general instructions as to the use of his product.

It is questionable, however, whether a general disclaimer, would exonerate a manufacturer to whom the purchasing employer had communicated his intention to use the commodity for a purpose other than that recommended. Indeed the value of such disclaimers may be doubted more generally.

(d) Labelling of articles and substances

One effect of these provisions is to encourage manufacturers and suppliers to stipulate fairly limited uses for their articles and substances in the hope that the burden might shift in part to the purchasing employer (under s2) to ensure that whatever he buys for his organization is correct for his requirements and is used only in accordance with the instructions supplied with it. The manufacturer himself might do well to ponder whether he is fulfilling his duty if he supplies goods in bulk with only a covering letter as to the restrictions he places on their use. Similarly, the supplier of machinery may do well to consider the extent to which the uses and misuses of the equipment should be printed upon the machine itself rather than supplied in accompanying literature.

Section 6(1)(c) (articles) and s6(5)(c) (substances) require only that there 'be available in connection with the use of the article (substance) for use at work adequate information'. In many situations, the Packaging and Labelling of Dangerous Substances Regulations 1978 will apply and it will be necessary, as a minimum requirement, to label the commodity in strict accordance with the requirements of these regulations; compliance with the regulations may not however be a complete discharge of that part of the s6 duty which requires the provision of information and the manufacturer or supplier of commodities which are covered by the regulations would do well to consider whether compliance with the regulations will discharge his general duties under the Act or whether particular information he has about the use to which his customer intends to put the commodity places some additional duty upon him. Also the general duty may well extend to circumstances where there are no specific labelling requirements. (For the labelling of toxic and hazardous substances in transit see The Dangerous Substances (Conveyance by Road or in Road Tankers and Tank Containers) Regulations.)

(e) Relationship between s2 and s6

Perhaps it should be mentioned that the employer who used a commodity for purposes other than those specified by the manufacturer will not automatically commit an offence; but he might well expect to have to discharge the burden of proving that the commodity was in fact safe for the purpose for which he used it and that he was therefore not in breach of his duty to his employees under s2. An interesting matter of technical evidence might well arise if an employer used something for a purpose or in a manner which the manufacturer specifically prohibited. In those circumstances the manufacturer's caution would protect him from liabilities under s6 but once again the employer's liabilities under s2 would not be clear. Yet another instance in which the issue would be likely to be determined by technical evidence could arise where the employer discovered a defect in the commodity which, in his opinion, made it unsafe in spite of the manufacturer's assurances to the contrary. Problems of this sort would be more likely to occur with machinery or equipment acquired as a long term investment. Here it would appear that the employer would be well advised to seek the advice of the Health and Safety Executive.

(f) Public liability

The duties arise only with articles or substances designed or intended for use (whether exclusively or not) by persons at work[11]. Once the duty has arisen, a reasonable interpretation of s6, in the light of the general purposes of the Act (s1(1)) is that it would also be owed to persons other than persons at work in appropriate cases[12]. If this view is correct, the designers and manufacturers of chemical plant for instance, would have a duty to the public in the vicinity of the chemical works to prevent explosion.

A further possibility is that these obligations will not be without influence upon the rights of the retail consumer, in appropriate situations. It seems doubtful whether a manufacturer or supplier could safely adopt one code of conduct for industrial users and another for domestic users; for even though there might be no penal sanctions in retail transactions for the domestic market, there could be civil liability. For example, if a paint sold for industrial use is labelled 'dangerous to the skin' could the manufacturer reasonably expect to escape liability for negligence to a retail customer who bought an unlabelled can in a do-it-yourself shop and suffered dermatitis as a result of using the paint to decorate his home? The fact that the Packaging and Labelling of Dangerous Substances Regulations apply equally to 'trade' and 'consumer' transactions is an indication of some intention to harmonize the statutory standards in these branches of the law. This will cover many situations, though not of course, the situations where s6 requires something in addition to the Regulations[13].

Generally the requirements of this section must be distinguished from the commercial considerations which will govern the contract under which the employer places an order for goods to be used by his employees at work. For commercial reasons the purchaser will wish to persuade the supplier to produce a commodity for him at the lowest possible price, while the supplier will wish to obtain the highest possible price for the commodity. Neither may be particularly anxious to pay the price required or undertake the effort necessary to ensure safety. The implications of this section, requiring safety standards from the manufacturer, may well turn out to be very great. It is highly probable that the section will be implemented by further regulations and it is likely that in the long run the effects upon the machinery and plant which is marketed for industrial use will be comparable to the effects of the vehicle construction regulations upon the car industry.

Duty not to interfere

(Section 8)

Section 8 provides that *no person* shall intentionally or recklessly interfere with or misuse anything provided in the interests of health, safety or welfare in pursuance of any of the relevant statutory provisions. The section applies the concept of s143 of the Factories Act more widely; whereas that section imposed a duty upon employed persons, the present section places that duty on the world at large. Since previous sections have identified the persons directly involved in activities associated with employment, this section would seem to be of particular relevance to the general public. It is likely to be especially important in premises which have a large population of visitors who are not within the category of persons at work, such as customers in a shop, students in a college or patients in a hospital. There is no reason why it should be confined only to members of the public and there could well be occasions upon which this section might be used against an employer, a manufacturer or an employee or, perhaps most likely, a self-employed person. It may well be that the duty imposed upon the self-employed under s3(2) – to conduct his *undertaking* in such a way as to ensure, so far as is reasonably practicable, that he and other persons are not exposed to risk – is not a particularly wide duty. If s3(2) were construed narrowly so that, for example, it did not cover the situation where a self-employed person takes staging from another's scaffolding for his own scaffolding, then s8 would appear to cover the situation.

5 The agencies (Sections 10 to 14, 55 to 60)

Introduction

Previous legislation relating to occupational health and safety was enforced by separate inspectorates; each of the major statutory codes had its own specialist and independent inspectorate attached to a particular Ministerial department. The defects of this piecemeal approach were many; for example, professional expertise was dissipated, there were demarcation disputes between inspectorates, on occasions there was a multiplicity of inspectorates concerned with one set of premises and there was insufficient concentration at Ministerial level on problems relating to health and safety. One of the major recommendations of the Robens Report was for the reform of the administration and enforcement system: an important object of the enabling Act was the establishment of a more efficient system of administration and enforcement.

The Commission

(Sections 10 to 14)
Among the first provisions of the Act to come into operation (October 1974) were the sections relating to the establishment of the Health and Safety Commission (HSC).

The Commission is a corporation consisting of a Chairman (the first Chairman was Bill Simpson) and not less than six, and not more than nine, other persons. The Commission must be a tripartite body with one third of its membership appointed after consultation with each of the organizations representing employers, employees and the local authorities and such other organizations, including professional bodies, as the Secretary of State considers it appropriate to consult.

The Commission is responsible to Parliament through those

departments of State (principally the Department of Employment) which are concerned with any aspects of occupational health and safety. While the Commission has wide powers, and is largely autonomous in the performance of its functions, it is ultimately subject to the control of the appropriate Secretaries of State.

It is the general duty of the Commission to do such things, and make such arrangements, as it considers appropriate for carrying out the general purposes of the Act, including the replacement of the earlier enactments and regulations by a system of regulations and approved codes of practice (see chapter 6) designed to maintain or improve standards of health and safety and welfare at work.

The duties of the Commission include assisting and encouraging people to carry out the general purposes of the Act, making arrangements for carrying out research, publishing, and encouraging others to publish, the results of research and providing training and information for these purposes[1].

The Commission is also required to make such arrangements as it considers appropriate to ensure that government departments, employers, employees, employers' organizations and unions are provided with an information and advisory service on the implications of the Act.

It is the duty of the Commission to submit proposals for regulations to the Secretary of State. It can also approve and issue codes of practice itself or approve codes of practice issued by others.

The Commission is empowered to investigate any accident or occurrence if it considers it necessary or expedient to investigate it.

The Executive

(Sections 10 and 11)

The Health and Safety Executive (HSE) is a corporate body consisting of three people. The Executive was created on 1 January 1975 and became the employer of most of the inspectors who had previously been appointed by virtue of the earlier safety codes, and who had been employed by many different ministries. Local authority inspectors employed to inspect offices and shops continue to be employed by their local authority but are required to work in close liaison with the Executive.

It is the duty of the Executive to exercise those of the Commission's functions as the latter delegates. Its primary function is to make adequate arrangements for the enforcement of the relevant statutory provisions except insofar as the Secretary of State has conferred the duty of enforcement upon some other authority.

At present the Executive operates mainly from Baynards House, Chepstow Place, London W2. Advice on the operation of the Act can be obtained from here or from the local offices of the Inspectorate.

Local authorities

In the past local authority inspectors had responsibility for offices, shops and railway premises by reason of the Offices, Shops and Railway Premises Act 1963. Now they retain responsibility for these premises under this Act and in addition have authority to enforce in them the provisions of the 1974 Act. The 1974 Act provides in s18 that the Secretary of State may by regulations make local authorities responsible for the enforcement of the relevant statutory provisions in other premises, and for transferring enforcement duties from the Executive to local authorities and *vice versa*. It follows that sections of the workforce, such as lecturers, teachers and nurses, who have been brought into the ambit of protective legislation for the first time, are within the province of the Executive until such time as special regulations confer jurisdiction over them onto local authorities. It seems likely that any new responsibilities allocated to local authorities will be in respect of 'non industrial activities'[2]. Until 1 June 1977 the financial straits in which local authorities found themselves caused them to resist even the imposition of the duty to enforce the 1974 Act in premises already within their responsibility. So far they have successfully resisted the allocation of further responsibilities to them in respect of 'new entrants'.

Other authorities

The Commission is empowered to make arrangements with any government department or person to perform on behalf of the

Commission or Executive any of the functions of the Commission or the Executive (s13(1)). When the 1974 Act was extended to cover offshore installations, the Health and Safety Executive delegated enforcement of occupational health and safety laws on oil and gas rigs in the North Sea to the Petroleum Engineering Division of the Department of Energy, the enforcement agency which had previously enforced the provisions of the Mineral Workings (Offshore Installations) Act 1971 offshore.

This power might also be used where there are overlaps between occupational safety and, say, road safety or public health. In these circumstances the Act empowers the Commission to reach agreement as to demarcation on enforcement with the Department of the Environment. The boundaries between the requirements of the Control of Pollution Act 1974 and the Road Traffic Acts and this Act can be considered in this way. The Department of the Environment has published a Code of Practice on the Carriage of Radioactive Materials by Road (under the Radioactive Substances Act 1948). This code specifically states it does not affect liability under such other legislation as the Nuclear Installations Act 1965, which is one of the relevant statutory provisions under the Health and Safety at Work Act. Similarly regulations concerning the carriage of hazardous and toxic substances are enforced by police on public roads and otherwise by the Health and Safety Executive.

The Inspectorate

The body of inspectors who enforce the 1974 Act and other relevant statutory provisions concerned with occupational health and safety consist in the main of the old inspectorates which had been set up under the principal protective employment codes, namely the Factories Act 1961, the Mines and Quarries Act 1954, the Offices, Shops and Railway Premises Act 1963 and the Agriculture (Safety, Health and Welfare Provisions) Act 1956, together with those enforcing less familiar statutes such as the Explosives Acts and the Nuclear Installations Acts.

When the Act became operative it was necessary to issue a large number of regulations to transfer the powers under the existing legislation to the newly formed inspectorate. The new inspectorate was given powers, to take effect from 1 January 1975, to enforce the existing statutory provisions; from 1 April 1975, when the general duties under the 1974 Act came into operation, the inspectorate had the task of enforcing these too. At present, therefore, the Inspectorate is charged with enforcement of both the old law (until it is replaced) and the new. It has been empowered since 1 January 1975 to employ its new enforcement tools in respect of the provisions of the pre 1974 legislation. Moreover, if an inspector institutes a prosecution for the commission of an offence under the pre 1974 legislation, the case may in many instances be triable[3] before a jury at the Crown Court; the offence will then be subject to the higher penalties which the 1974 Act permits in the case of a conviction following a trial upon indictment.

In England and Wales, only an inspector may institute a prosecution for breach of any of the relevant statutory provisions, except with the consent of the Director of Public Prosecutions. In Scotland prosecutions are normally conducted through a Procurator Fiscal.

The Employment Medical Advisory Service (EMAS)

(Sections 55 to 60)
The EMAS is maintained under the 1974 Act for the purpose of informing and advising government bodies concerned with the health of employed persons or of persons seeking, or training, for employment, and also of informing those persons concerning the safeguarding and improvement of their health.

The responsibility for maintaining the service is placed on the Secretary of State, but it is clearly envisaged that the practical task of maintaining the service will be carried out by the Commission.

6 The safety standards (Sections 15, 16, 50 and Schedule 3)

Introduction

The Robens philosophy, reflected in the Act, was to impose a general requirement to do all that is reasonably practicable for achieving health and safety at work. In time, as earlier legislation is repealed to make way for revised standards of protection, the only remaining statutory provisions for health and safety will be the general duties and regulations made under the 1974 Act.

The statutory standards relating to health and safety at work will then be found in the regulations, which will assume far more importance in this field in the future than they have had in the past. Their effect will be reinforced by codes of practice; a system with which most managers were unfamiliar before 1974.

Power to make regulations

(Sections 15 and 50)

The Secretary of State is given power under s15 to make regulations, to be known as health and safety regulations, for any of the general purposes of the Act.

The Act thus gives the Secretary of State very wide powers indeed to make regulations, including provision for the health and welfare of persons at work and the protection of persons other than persons at work against risks to health or safety arising out of or in connection with the activities of persons at work. The general purposes of the Act also specifically include the replacement of existing legislation as well as the regulations, orders and other instruments made under them, by revised regulations and codes of practice. An indication of the type of subject on which regulations can be made under these powers can be found by studying

Schedule 3 of the Act; but this list is only indicative and should not be regarded as exhaustive.

In s50, the Commission is placed under a duty to submit proposals to the Secretary of State for making regulations: the Secretary of State may, alternatively, initiate proposals himself but, if so, he is required to consult the Commission before exercising his power to convert the proposals into regulations. When either the Commission or the Secretary of State propose regulations, the proposer is required to consult appropriate bodies, that is representatives of employers, employees and, possibly, local authorities and other organizations. Since the passage of the Act the Commission has initiated consultation on a number of subjects and substantial numbers of regulations have resulted from these consultations.

Scope of the regulations

The scope of health and safety regulations is set out in detail in s15(3)(4)(5) and (6). These provisions deserve study because of the light they throw on the likely form that regulations will take. For example, regulations may:

(a) repeal or modify existing statutory provisions. This is in accordance with the general purposes of the Act to replace the existing legislation with more appropriate modern provisions

(b) exclude or modify in relation to any specific class of case any of the provisions of the sections relating to general duties, or of any of the existing statutory provisions. This enables regulations to be made that are either stricter or less strict than a general duty in a specific class of case (for example to re-enact the present 'strict' fencing of machinery obligations in the Factories Act)

(c) specify the authority responsible for the enforcement of the relevant statutory provisions. In most cases, one may expect the Executive to be the enforcement authority but it might be a local authority or other authority

(d) provide that regulations may specify the persons who, in the event of a contravention of a requirement or prohibition imposed by or under the regulations, are to be guilty of an offence. This provision would enable regulations to place duties upon individuals in the management hierarchy, and possibly upon employed persons. It will have been observed that the general duties are imposed upon 'principals' such as employers, controllers of premises, manufacturers and suppliers. Section 15(6) enables regulations to impose legal obligations upon any suitable person and thus, if it is considered appropriate, impose a pattern of direct responsibility within an organization, such as is found in the existing Mines and Quarries Act 1954

(e) provide for specific defences to be available in proceedings under the relevant statutory provisions. The general duties are, in the main, subject to the qualification (which in s40 is regarded as a defence) of 'reasonable practicability'. But defences under the regulations might take any form deemed appropriate and might include a 'third party' defence similar to that found in s161 of the Factories Act or s150 of the Mines and Quarries Act

(f) restrict the penalties which may be imposed for particular offences. In practice, this may be either a stipulation as to maximum fines for the offences created by the regulations, or else a provision to exclude trial upon indictment for breach of particular regulations. It is also provided that regulations may exclude proceedings upon indictment in relation to offences under the general duties, the existing statutory provisions or health and safety regulations.

It is interesting to note that regulations made since 1974 have in a number of instances been of general application with an integrating function; that is replacing various requirements contained in the earlier statutory provisions with a requirement applicable to all work situations (eg The Notification of Accidents and Dangerous Occurrences Regulations and First Aid at Work Regulations). In the Control of Lead at Work Regulations, a number of specific requirements are replaced by general duties to safeguard employees from the hazards of lead.

Subject matter of regulations

(Schedule 3)
Schedule 3 of the Act sets out a formidable list of matters which may be the subject matter of the health and safety regulations. As

might be expected, some part of this Schedule is concerned with authorizing the making of regulations relating to standards for plant and machinery, and to the maintenance of a safe working environment and safe systems for production and transportation of substances. Regulations made on these matters are likely to be within the traditional approaches to safety problems and ensuring their observance may well be exclusively within the province of the technical staff. However, the Schedule gives a clear indication that the Act makes provision for regulations in provinces which are less traditional; a very large part of the Schedule is concerned with authorizing regulations concerning the activities and competence of persons at work. Therefore, regulations may well appear which impinge upon the personnel manager's work relating both to the selection of employees for particular tasks and also to the training of persons once they are in post. Regulations may relate to the provision of welfare facilities and also to management systems. It is worthwhile noting some of the matters upon which regulations may be made according to the provisions of the Third Schedule, and considering their special significance for the management of human resources.

(a) Licensing and registration

Para 4(1) enables regulations to be made prohibiting the carrying on of any specific activity or the doing of any specified thing except under the authority and in accordance with the terms and conditions of a licence; or except with the consent or approval of a specified authority. The use of licensing is already well known in some branches of industrial activity, but it may well be that under this Act there will be an increased emphasis on the technique, which may be used in respect of persons as well as premises or processes. For example crane drivers and fork lift truck drivers might well have to be licensed in the foreseeable future.

Para 5 permits regulations requiring any person, premises or thing to be registered in any specified circumstances or as a condition of the carrying on of any specified activity or the doing of any specified thing.

Para 6(1) enables regulations requiring, in specified circumstances, the appointment (whether in a specified capacity or not) of persons with specified qualifications or experience both to perform specified functions and imposing duties or conferring powers on persons appointed (whether in pursuance of the regulations or not) to perform specified functions[1].

Para 6(2) enables regulations restricting the performance of specified functions to persons possessing specified qualifications or experience[2].

Clearly all these provisions relate to the recognition of the importance of installing and maintaining safe systems in which responsible and qualified persons are appointed to undertake specific tasks. Regulations with these objectives are of significance to general management both in the selection of employees and also to the personnel manager in framing the terms of contracts of employment for those to whom such responsibilities are allocated. It seems likely that regulations will require persons employed in the management of specified dangerous occupations to possess some appropriate certificates of competence before they are deemed qualified for a purpose envisaged in the regulations[3]. Where a regulation requires only registration, the principal purpose of the requirement is likely to be merely to identify responsibilities; it may not necessarily follow that, to be registered for a specific purpose, an individual must possess formal qualifications or for that matter particular competence.

It is significant that the Act imposes the most severe of its penalties – namely two years imprisonment – for, *inter alia*, unlawfully operating without a licence or for contravening the terms of a licence. It seems likely that there may be instances in which regulations may require full licensing as a method of ensuring standards, and also of specifically allocating responsibilities[4].

In future, the safety officer employed in hazardous industries may require specific qualifications or, indeed, need to be licensed. It is conceivable that the personnel manager himself might be affected by some equivalent regulation relating to his responsibilities and to his competence, especially if training comes within his area of responsibility.

(b) Health

Para 8(1) enables regulations requiring the making of arrangements for securing the health of persons at work or other persons, including the arrangements for medical examinations and health surveys.

Para 8(10) enables regulations for securing the provision of specified welfare facilities for persons at work, including in particular such matters as adequate water supply, sanitary conveniences, washing and bathing facilities, ambulance and first aid arrangements, cloakroom accommodation, sitting facilities and refreshment facilities.

Regulations on monitoring the health of persons at work, individually, by medical examinations, are likely to come within the province of the personnel manager. The failure on the part of an employee to submit himself to statutory medical examinations could be regarded as a disciplinary offence independently of any express contractual agreement to undergo such examinations.

Keeping of records and the organization of periodical medical examinations may also concern the personnel manager.

Similarly, in many cases (since 'welfare' is traditionally within the province of the personnel manager), it may well be the personnel manager's function to ensure that the company complies with the standards imposed by regulations for these matters (eg compliance with the medical suspension provisions of the Employment Protection (Consolidation) Act).

(c) Instruction and training

Para 14 enables regulations to be made 'imposing requirements with respect to the instruction, training and supervision of persons at work'.

Such regulations would be likely to be closely linked to regulations requiring competent persons for particular jobs. These regulations could, however, be wider than the formal requirements for licensing and the appointment of competent persons. Regulations relating to training could lay down minimum standards and restrict the concept of qualification to these standards, without necessarily requiring that every practitioner now in post possessed the qualifications – for an interim period at any rate.

Regulatory standards specifying competence could assist personnel managers and training officers to identify appropriate courses to which employees of particular grades might be assigned, but it is also quite likely that such regulations will impose duties to ensure that by those means, or by on-the-job training, operatives are competent for the job and are adequately supervised both in their training period and thereafter[5].

Finally, training might be required under these regulations at management as well as at operative and skilled manual levels.

(d) Information

Para 15(1) enables regulations requiring, in specified circumstances, specified matters to be notified in a specified manner to specified persons.

This paragraph is phrased so widely that it could be implemented to relate to the giving of information to persons within the organization, for example, the giving of information to employed persons concerning hazards in their employment; or to the giving of information to persons outside the organization, for example, to the provision of statistical returns to the Executive or the Commission[6]. It could also relate to providing information to the public in the case, for example, of neighbourhood risks.

Legal status of regulations

The technique of relying heavily upon regulations to implement a parent Act is widely used in respect to industrial activity and is therefore familiar to most managers. It cannot, however, be stressed too strongly that regulations on occupational health and safety are likely to assume a new importance under this Act, because a far greater proportion of safety law than ever before is ultimately likely to be contained in regulations rather than in an Act of Parliament. This is so because of the breadth of the parent Act itself, and also because of the wide nature of enabling provisions made both in Schedule 3 and in the body of the Act. Moreover, because of the overriding importance of the general purposes section of the Act, the provisions of that Schedule relating to the subject matter of regulations may be little more than a list of examples of the regulation-making power. It therefore seems unlikely that a lawyer will have much opportunity to challenge the validity of regulations on the ground that the Secretary of State has acted beyond the power delegated to him by Parliament to make regulations. It might be, of course, that the Secretary of State would act unlawfully if he made regulations without going through the forms of consultation required by the Act.

Regulations are made by statutory instruments and Parliament has the power to annul them. They are also subject to the supervisory jurisdiction of the law courts.

Regulations are enforced in exactly the same way as the Act (for enforcement *see* chapter 6). It may not always be clear, whether compliance with a regulation made to spell out a particular aspect of a general duty will necessarily amount to compliance with that general duty in its entirety in respect of the situation covered by that Regulation. Much will doubtless depend on the wording of the regulation, which may expressly stipulate that compliance with the regulation itself will, in the circumstances covered by the regulation, amount to doing all that is reasonably practicable for the health and safety of another.

As a matter of general principle, it is difficult to maintain that compliance with a specific regulation would always amount to compliance with the general duty to do all that is reasonably necessary to ensure the safety of employees; for example regulations may require eye protection to be provided to employees in particular circumstances, but an employer may have failed to provide sufficient information, instruction, training and supervision to ensure that the equipment provided is properly used on all necessary occasions. So it seems clear that mere compliance with regulations on the provision of eye protection would not, even in respect of the protection of the employee's sight, amount to full compliance with the general duty imposed by s2.

It is also possible that the regulations, for technical reasons, are not strictly applicable (for example, they might apply only to factory premises and the instance in question might exist in a hospital) and might well be utilized by the inspectorate although only of evidential value, as indicating the standards with which an employer might be expected to comply if he had a proper regard for the health and safety of his employees.

Codes of practice

Section 16 of the Act empowers the Commission to approve codes of practice in order to provide guidance on the requirements of any provision of s2 to s7 of the Act, or of safety regulations or of any of the earlier statutory provisions. The Commission may produce codes on its own initiative, adopt codes prepared by other bodies or give a stamp of approval to codes which others have issued.

The Commission cannot approve a code of practice without the consent of the Secretary of State, and before seeking his consent the Commission is required to consult appropriate bodies, which includes employers and employee representatives. The Commission may equally approve a revision of, or withdraw its approval from, a code provided that the same procedure, including consultation, is followed.

A type of code of practice, which give guidance in matters relating to health and safety at work, had been produced by departments of State before the 1974 Act and were, and still are, available from HMSO. A well known example was the *Code of Practice for Reducing the Exposure of Employed Persons to Noise*, published by the Department of Employment in 1972.

Such codes were intended to give practical guidance on the safe conduct of industrial activity. They had no binding force; but it was hoped that common sense would encourage their adoption. There was some incentive to adopt them for safety purposes, in that the employer who failed to do so might find his failure made him vulnerable in civil litigation if an employee claimed compensation from him for personal injury. In these circumstances, the employer who had failed to conduct his business in accordance with a well publicized code might have difficulties in persuading the court that he had exercised reasonable care for the safety of the plaintiff. Because of this background the concept of codes of practice is familiar, but the Act gives such codes vastly increased importance in the criminal courts for the enforcement of safety law.

Scope of codes

Those codes of practice which were published prior to the Act suggested that a code was likely to be especially applicable when the introduction of safe systems was sought to minimize the exposure of persons to dangerous situations. This would apply with particular force where there were health risks: codes might well then be adopted to give guidance on safe levels of exposure, methods of measuring the extent of the hazard, precautions which should be taken to protect employees exposed to the hazard and methods of monitoring the health of employees who are exposed.

Approved codes published since the Act have in fact followed this pattern (*see* the Code of Practice on Control of Lead at Work).

Approved codes of practice are frequently published simultaneously with regulations. The requirements contained in the regulations, and often the codes themselves, are in turn accompanied by guidance notes giving the inspectorate's view of how the law should be applied (for example The Safety Representatives and Safety Committees Regulations are supplemented by both a code of practice and guidance notes).

A code of practice may be relevant in consultation matters. The use of codes of practice in industrial relations was established by the Industrial Relations Act 1971 and there are now a number of codes in existence under other employment statutes, for example the code on Disciplinary Rules and Procedures. This code is relevant to safety in as much as it relates to procedures for dismissals for misconduct which could occur where safety rules are not observed. That code should not be confused with codes of practice under this Act, because the breach of an industrial relations code has no criminal consequences.

Legal status of codes under the Act

Codes of practice relevant to health and safety at work are now of importance in the enforcement of accident prevention law as well as in accident compensation law. Under s17 of the Act it is provided that, while any person's failure to observe any provision of an approved code of practice shall not, of itself, render him liable to either criminal or civil proceedings, nevertheless where a person is alleged to have broken a general duty, regulation or other relevant statutory provision, the fact that the accused has failed to observe a relevant code of practice may be taken as conclusive evidence of his failure to do all that is reasonably practicable to ensure the health and safety of those at work, unless the court is satisfied that the defendant complied with his obligations in some other way.

Additionally codes of practice may be used in enforcement proceedings before industrial tribunals, under s23. The general enforcement of the Act is discussed in chapter 7.

7 Enforcement of the Act

The role of inspectors

(Sections 18 and 19)
The Act places the main burden of enforcing the relevant statutory provisions upon the Executive, in cooperation with local authorities and, where appropriate, other authorities. The day to day work of enforcement devolves, as in the past, upon specialist inspectors. The right to initiate proceedings under any of the relevant statutory provisions is virtually limited to the inspectors by s39.

The Act requires the appointment of an inspector to be in writing (s19) and the document in which he is appointed must specify which of the powers made available to inspectors generally under the relevant statutory provisions shall be exercised by him in the course of his duties. The appointment of an inspector is not invalidated by any defect in the instrument creating the HSE[1].

Inspectors' general powers

(Sections 20 and 25)
The powers of the inspectors are set out in s20. They are far more extensive than those granted under any of the earlier statutory enactments, including as they do all those found in any of these enactments, together with some remarkable additional powers under the Act. Thus an inspector may:

(a) at any reasonable time, enter any premises which he has reason to believe it is necessary for him to enter for the purpose of enforcing the relevant statutory provisions. This is a considerably wider power than that contained in s146(1)(a) of the Factories Act where a power to enter was given only in respect of factory premises in which the inspector had reason to believe persons to be employed. The Act, for example, extends the inspector's jurisdiction to premises in which the only workers are partners of a firm, or to premises which are used entirely as a warehouse where no persons are employed. Some extension of the old power was clearly necessary, as the Factories Act only granted protection to employed persons in factory premises, whereas the present Act gives general coverage to persons at work. The 1974 Act goes rather further than this, and does not even require the likelihood of persons being at work in premises to justify the inspector's right of entry to them

(b) make an examination and investigation on the premises. For this purpose he may take with him any equipment or materials he needs. He may take measurements, photographs and recordings. He may take samples, dismantle equipment and test substances. Before exercising these powers he should consult such persons as appear to him appropriate to ascertain what dangers, if any, there may be in carrying out his intention. He must exercise his powers to dismantle and test (if requested to do so) in the presence of the person who is present in, and has responsibilities for, the premises: the manager in charge

(c) order that any part of the premises or anything on the premises be left undisturbed for as long as is necessary to enable an examination or investigation. The object is no doubt to facilitate the investigation of an accident or dangerous occurrence, but it should not be forgotten that it might entail bringing production to a halt in some instances.

He may take possession of any article or substance for the purpose of examination and to ensure its availability as evidence in any criminal proceedings, or in any proceedings in relation to a notice served by an inspector. Where he takes possession of any article or substance found on any premises he must leave there, with a responsible person (or if that is not practicable, fixed in a prominent position), a notice giving particulars of that article or substance and stating that he has taken possession of it. Before taking possession of a substance he is required, if practicable, to give a sample thereof to a responsible person acting for the firm

(d) where he has reasonable cause to believe that an article or substance is a cause of imminent danger of serious personal injury, he is empowered by s25 to seize it and render it harmless. In so doing the inspector is bound to follow the procedure set out in that section

(e) require any person whom he has reasonable cause to believe to be able to give any information to answer such questions as he sees fit to ask and to sign a declaration as to the truth of his answers. But no answer given by a person shall be admissible in evidence against that person in any proceedings. This power is wider than the comparable Factories Act provisions which were restricted (except for a general power to question any person as to the identity of the occupier of particular premises) to the questioning of persons supposed by the inspector to be employees in particular premises

(f) require any person to give him such help and assistance on any matter within that person's control or in relation to which that person has responsibilities as are necessary to enable the inspector to carry out his statutory powers

(g) require the production of any books or documents which it is necessary for him to see for the exercise of his powers. This again is wider than the Factories Act, which only permitted the inspector to require the production of such registers, certificates, notices and documents as were kept in compliance with the Act. Under his present powers the inspector might reasonably require the production of correspondence, eg between an employer and the supplier of a substance for use at his workplace. But no person can be compelled to produce any document which he could withhold from discovery in legal proceedings on the ground of legal privilege.

In a sense this information gathering is central to the inspector's function; it is buttressed by an exceptionally wide power granted to the Commission itself in s27. With the consent of the Secretary of State for Employment, the Commission may require any person or body to provide any information deemed necessary for the discharge of the functions of the Commission itself or of any of the enforcing authorities (which will be, in most instances, the Inspectorate or the local authorities). The only limitation on the information that may be called for is the 'general purposes' of the Act, namely health, safety and welfare of persons at work, or the health and safety of persons affected thereby. This power is seen to be particularly significant if related to the inspector's duty under s28(8) to give employees factual information related to the premises at which they are employed. The inspector is also required to provide factual information to parties in civil proceedings (s28(9). (This requirement was introduced by the Employment Protection Act 1975 (s116 and Schedule 15).)

Inspectors' enforcement powers

(Sections 21, 22, 23 and 24)
The enforcement powers created by the Act were novel in two respects: first, the concept of notices served by the inspector was quite different from any of the provisions of earlier legislation and, secondly, the criminal sanctions which a court can enforce upon a wrongdoer were much more severe than had been permitted under previous occupational safety legislation.

In the event of investigation disclosing a situation which the inspector considers unsatisfactory, he may serve a notice requiring the situation to be remedied and in addition, or alternatively, he may institute a prosecution alleging a specific breach of a relevant statutory provision.

(a) Relationship between notices and legal proceedings

The notice is an instruction from an inspector to a person to remedy a situation which, in the opinion of the inspector, is illegal or unsafe. It requires immediate action in one form or another from the person upon whom it is served. He must either take steps to comply with the notice, within the time specified in it, or else challenge its validity in an appeal before an industrial tribunal; if he merely lets the time run out against him, he commits a criminal offence. Effectively, therefore, the provisions relating to the inspector's powers to serve notices make it a criminal offence to fail to do what the inspector requests, provided he makes his request formally in the appropriate notice.

This is a very different situation from that which existed before the relevant sections of the Act came into force. Up to that time the inspector could only advise a person to rectify what in his opinion was a breach of the law. It was not an offence to fail to comply with the inspector's advice and the only mechanism for establishing whether the advice of the inspector was correct was for him to institute a prosecution in the magistrates' court, alleging a contravention of the law. In court the magistrates would have to deal with both the well informed defendant who honestly disputed the validity of the inspector's requirement, and the hardened offender who merely wished to postpone compliance with a request whose validity he did not seriously challenge.

The merit of the present system is that it enables a genuine dispute on the reasonableness of an inspector's requirement to be discussed in an industrial tribunal as a technical argument conducted by the inspector and the other party, in the presence of a tribunal which may be specially constituted to deal with technical arguments. The criminal courts can now concentrate upon the real offenders, including the persons who allow time to run out under a notice without complying with it, and those who, having unsuccessfully appealed, afterwards decline to accept that the industrial tribunal has upheld the inspector's opinion.

There are two forms of notice, the improvement notice and the prohibition notice.

(b) Improvement notices

(Section 21)

An inspector may serve an improvement notice on a person when he is of the opinion that the person is contravening one or more statutory provisions. The notice will state the inspector's opinion, specify the provisions which he believes to be broken, give the reasons why he believes this, and require the employer or the appropriate transgressor to remedy the situation within a specified period. An improvement notice may be served in respect of breaches of earlier legislation such as the Factories Act, as well as of the duties and regulations created by virtue of the 1974 Act.

The Act requires an inspector who is serving a notice concerning premises he has visited to notify the employees working there of his intention to serve the notice.

Since the notice must be served upon the person who is, in the opinion of the inspector, breaking a duty imposed upon him by the law, the selection of a person to receive the notice will depend upon the nature of the duty which the inspector is seeking to enforce. If, for example, he serves the notice because he believes that the inside walls of a factory are not kept clean in compliance with s1 of the Factories Act 1961, he will serve the notice on the person who is technically the occupier of the factory, since he is the person who is under a duty to comply with the section in question. On the other hand, if his complaint relates to a breach of the general duty imposed upon the employer under s2 of the Health and Safety at Work Act 1974, he will serve the notice upon the employer. Regulations may impose duties on employees; in such cases, the notice could be served upon the appropriate employee. Indeed, notices have been served in respect of a breach of the general duty imposed upon employees under the Act itself.

It follows that an improvement notice relating to a contravention of the Factories Act 1961 would not be served upon the managers of the factory in their personal capacity. However, a notice relating to the Mines and Quarries Act 1954 or to health and safety regulations (either of which might impose duties upon management) might well be served personally in the future.

It is unlikely that a notice would be served directly on a personnel manager, since he is not the person in contravention of statutory duties. It could well be that he is normally the person to whom responsibility was delegated to see that an order was complied with. This might happen, for example, if a notice were served in respect of consultation machinery within the organization, or if it related to unsatisfactory systems for training employees. In such cases, it would be possible, although unlikely, that the personnel manager might be implicated personally if the organization for which he worked were prosecuted for failure to observe the notice. This would be likely only if the personnel manager were personally at fault in that he had negligently or wilfully failed to carry out the responsibilities delegated to him. In such circumstances, it would seem that his liability would be likely to be additional to, rather than in place of, the liability of the organization itself.

(c) Prohibition notices

(Section 22)

An inspector may serve a prohibition notice upon a person who is in control of activities which, in the inspector's opinion, involve a risk of serious personal injury. The prohibition notice will state the inspector's opinion that there is a risk of this nature, specify the matters which in his opinion give rise to the risk and direct that the activities to which the notice relates shall not be carried on by or under the control of the person on whom the notice is served unless the matters specified in the notice have been remedied.

The inspector may well grant time for the remedying of the situation, but the Act provides that the notice shall take immediate effect if the inspector is of the opinion that the risk of personal injury is 'imminent'.

It will be noted that the prohibition notice, unlike the improvement notice, may be served upon a person who is not in breach of any specific requirement of the relevant statutory requirements. Although prohibition notices cannot be served except in respect of activities referred to in the general purposes of the Act, they do not have to relate to a specific breach of the law. These notices may well be served in respect of an unsafe situation rather than of a specific offence; although the inspector may, if he thinks fit, allege both an unsafe condition and a specific breach of the law in the same notice.

The prohibition notice is served upon the person in control of the particular situation both because technically there may be no breach of a specific statutory duty, and also because the person in control of the situation is the person best able to act reasonably rapidly to remedy it. It is not entirely clear who may be deemed to be in 'control', but it is suggested that it would be unlikely to be the operative or even the first line supervisor; it would more likely be the works manager, rather than the employer. To serve the employer would in most cases mean following the time-consuming procedure of serving notice on a corporate body at its registered offices, as required by s46, and time is of the essence in the serving of prohibition notices.

The penalty for contravening a requirement of a prohibition notice is of the severest kind permitted under the Act, namely up to two years' imprisonment, an unlimited fine or both. It is arguable that the wording of the penalty clause is sufficiently wide to allow someone other than the person upon whom the notice was originally served to be liable for contravention of its requirements, such as a member of the management chain[2]. Thus the notice might require the works manager to cease to use an obsolete and dangerous machine and the works manager might duly relegate the machine to the scrap-heap. A foreman might then interfere and put the machine back into use. In these circumstances it is possible that both the works manager and the foreman might be in contravention of the notice, if the foreman was aware that it had been served in respect of the machine. The works manager might be liable for failing to take adequate steps to comply with the notice and the foreman for wilfully contravening it.

(d) Complying with notices

(Section 23)
A notice may (but need not) include directions on the measures to be taken to remedy the situation to which it refers. It is provided that such directions may be framed by reference to a code of practice and may be framed in such a way as to give a choice between different ways of remedying the situation.

(e) Appeals against improvement and prohibition notices

(Section 24)
Section 24 provides a right of appeal to an industrial tribunal against both forms of notice. Regulations[3] have been made providing that appeals may be lodged within three weeks of the date of serving the notice. The notice of appeal must be lodged with the Central Office of Tribunals.

The operation of an improvement notice is suspended by lodging an appeal, until such time as the appeal is withdrawn or disposed of by the tribunal. In the case of a prohibition notice, the bringing of an appeal does not have the same effect. In this case only the tribunal itself is empowered to order suspension, either while hearing the appeal or at a specially convened hearing to deal with suspension as a preliminary point.

The tribunal may cancel, affirm or modify the notice as it thinks fit. The Act gives no further guidance to the tribunals, so we must assume that a wide discretion is bestowed upon them. For example, a tribunal may seek to balance the risk of injury to employees against the disruption to a company's business resulting from a prohibition order[4]; or in other words seek to achieve a balance between the quantum of the risk to employees on the one hand and the sacrifice involved in the measures necessary for averting risk on the other[5].

The tribunal would clearly cancel a notice if it were not satisfied that the law had been infringed, or that a risk of personal injury or health hazard existed[6]. Additionally, a notice might be cancelled, even though an inspector acted within his powers, if the tribunal considered that he had over-reacted to the situation by exercising his power in issuing a notice.

Another way in which a notice might be modified would be by extending the time limits for compliance with it, or by varying the measures to be taken to remedy a contravention. The tribunal is unlikely to be influenced by pleading that the financial position of the party on whom a notice has been served is such that he cannot afford to comply with it; to allow such pleading would give an unfair advantage to the appellant over his business competitors who did comply with their statutory obligations, incurring expenses (and possibly charging higher prices) as a consequence of doing so[7].

Proceedings at an industrial tribunal are informal, and the emphasis is upon the realities[8] of the safety problems which lead inspectors to issue notices, rather than the legal issues with which courts are normally concerned.

The tribunal is empowered to review its own decisions on a number of grounds, for example where new evidence becomes available[9]. Although there is no statutory provision for appeal, where the decision of the tribunal is arguably 'irregular' or 'improper' a right of petition may lie to a Divisional Court of the Queen's Bench Division; the Divisional Court has already accepted jurisdiction in such an appeal case[10].

Prosecution of offences

(Section 33 and 34)

As an alternative to, or in addition to, issuing a notice, an inspector may decide to initiate a prosecution for a contravention of any of the relevant statutory provisions.

A contravention of any of the relevant statutory provisions is, by virtue of s33, a criminal offence. Thus it is an offence to contravene the general duties, health and safety regulations or the terms of an inspector's notice. It is still an offence to contravene any of the provisions of the existing statutory provisions, so long as they remain in force. It is not an offence to act contrary to the recommendation of a code of practice, though ignoring a code could be used as evidence of the breach of a duty. Where a contravention has been established primary responsibility is imposed upon 'persons' – a term which will probably include, for this purpose, employers, occupiers, controllers of premises, members of the management team, shopfloor employees, visiting workers, and, in certain limited cases (eg s8), members of the public.

The width of this provision may be contrasted with many of the earlier statutory provisions under which primary offenders were narrowly defined (eg the Factories Act 1961, under which the primary offender was almost invariably the occupier). As the earlier statutory provisions become replaced by health and safety regulations the pattern of responsibility for all health and safety offences will gradually take the form laid down in s33.

Classes of offences under the Act

Offences under the Act fall into two categories: summary offences and offences which may be tried summarily or on indictment. The first category includes the less serious and more routine types of offence that do not directly endanger persons at work. These offences are triable in magistrates' courts. The second category includes *inter alia* failure to obey an inspector's notice, contravention of the general duties under the Act, regulations involving some degree of 'endangering' of persons at work, and also many offences under the earlier statutory provisions. The inclusion of the earlier statutory provisions gives rise to an anomalous situation; for the time being a failure to safeguard machinery in a factory, for example, remains an offence with the pattern of criminal responsibility laid down under the old law; but the penalty is that laid down under the Act. These offences may be triable either summarily or upon indictment, (ie either in a magistrates' court or in the Crown Court before a jury).

Cases which may be tried only summarily present few problems. Where offences are triable either on indictment or summarily, a magistrates' court dealing with the case will consider whether trial should be summary or upon indictment. If they decide the latter their decision is final: if the former it appears that the accused still has the right to insist upon jury trial if he so desires. It does mean, however, that where the prosecutor asks for summary trial, the magistrates have power, should they consider the matter a serious one, to require the case to be tried upon indictment[11].

Effect of trial by jury

The effect of making offences in relation to occupational health and safety triable at the Crown Court before a jury, with heavier penalties for the guilty, clearly makes it much more serious to contravene statutory rules relating to health and safety than it has previously been. The machinery enabling the service of notices as the first stage in enforcement may well explain why the mechanism of trial by jury is rarely invoked except in cases where there has been flagrant disregard for safety, or involving major hazards, or heavy loss of life. Thus it is only in the most serious cases that trial is before a jury. It is important to remember that a person is most unlikely to be convicted in a case which is sufficiently serious for it to be tried upon indictment, unless the court is satisfied that he not only committed a wrongful act but also that he intended to do so and was therefore deserving of punishment. Moreover, the courts have shown that the test of whether a man intended to commit a crime is to be assessed purely subjectively and not by asking what other people would have supposed him to have intended. Although it is possible that individuals may be convicted of offences, in which the sentence could be imprisonment, it would appear that this heavy penalty is not likely to be imposed except in the most flagrant cases of personal and wilful disregard for safety.

Penalties under the Act

(Section 33)
Penalties under the Act apply to all offences committed either under the Act or committed since 1 January 1975 under the earlier statutory provisions.

The maximum penalty that may be inflicted in any circumstances in a magistrates' court under the Act is £1,000, as a result of the Criminal Law Act 1977 (amending s33 of the 1974 Act). Similarly the normal punishment that the Crown Court can inflict following conviction is a fine, but this may be of any amount that the court thinks just.

In a limited class of case, the Crown Court may inflict up to two years imprisonment, either instead of, or in addition to, a fine of any amount. This class of case includes contravening a licensing requirement, disobeying a prohibition notice, or the unauthorized disclosure of information obtained by the Commission, the Executive, or other enforcing agencies. Thus no possibilities arise for the penalty of imprisonment being inflicted in cases of substantive breaches of the general duties, or of the regulations made under the Act. It must also be emphasized that in no circumstances are magistrates' courts empowered to impose a penalty of imprisonment for breaches of the Act that come within their jurisdiction.

Defences

(Sections 36 and 40)
At first sight there is a conspicuous absence of defences under the Act. On close observation it is possible to identify two sections which contain provisions which are akin to defences.

Section 36 provides that where an offence by the person deemed responsible is due to the 'act or default' of some other person, the latter person may be charged with the offence, whether or not proceedings are taken against the primary offender.

This section appears to be similar in intention to the former s160 of the Factories Act 1961, but it is significant that s36 speaks of the 'offence' while s160 referred to 'an offence' and to 'the like fine'.

The likely explanation for the difference in terminology is that s36 now makes provision for a situation in which an inspector may wish to take proceedings against the 'other person', as an alternative rather than in addition to proceeding against the primary offender. This interpretation is supported by the wording of s36(1) which specifies that a person may be charged and convicted of the offence whether or not proceedings are taken against the named person. It appears that this discretion will be exercised by the Executive and not by a court. It is perhaps misleading to describe this as a defence. It appears to be a basis for prosecution rather than a plea which the accused may raise at his trial in order to escape liability.

This view may be supported by contrasting the language of the section with that of the former s161 of the Factories Act, which provided for a 'third party procedure' in which a court was empowered to identify the actual offender against the provision in question. In this case the person primarily liable could use the provision to exempt himself from liability.

If this distinction between the two statutes is correct an ostensible primary offender under the Act of 1974, seeking to avoid proceedings, should make his representation to the inspector and not keep the evidence in reserve until his appearance before a court.

On the other hand, it appears feasible, under s36, that the inspector may exercise his power to take proceedings both against the primary offender and the other person, for example in a case in which there was evidence of a failure to give leadership on the part of the primary offender.

It is at least arguable that s40 provides, or at least includes reference to, a further defence. Certain requirements of the Act (eg the general duties) and the earlier statutory provisions are qualified, by reference to practicability, reasonable practicability, use of the best practicable means and so on. However this relates only to contraventions of provisions which expressly incorporate the phrase and has no relevance to provisions which do not do so[12]. This section has been discussed (*see* chapter 3) when the general duties were examined.

The section makes it clear that whenever these qualifications are found in a requirement, they will serve as defences in criminal proceedings under the Act or earlier statutory provisions, providing the accused can fully prove that it was not practicable, or reasonably practicable etc, to do more than was in fact done.

The health and safety regulations may contain their own defences; under s15(6)(a) and (b) they may specify the persons to be primarily responsible for breaches, any other person upon whom criminal responsibility may be placed and any specified defences which are available to them.

Application to the Crown

Section 48 states that the Act binds the Crown but, at the same time the Crown may not be prosecuted or served with a prohibition or improvement notice. The Act, however, appears to intend that Crown employees should enjoy no different treatment from that accorded to the employees of other organizations. Crown premises may be inspected by the Health and Safety Executive's inspectors.

It is a matter of controversy whether the Crown should enjoy this exemption; it must be borne in mind that the Crown is a very large employer, so that, for example, many industrial undertakings, the prison service and the National Health Service, all enjoy this exemption. The Health and Safety Executive has devised a special notice procedure for informing the Crown when it is in breach of its duties and has so far declined to enforce the Act against Crown employees.

Duties and liabilities of inspectors

Although the Act sets out in great detail the *powers* of the inspectors appointed under it, it says little about their *duties*.

The general law on the duties, as opposed to powers, of inspectors is far from clear. Obviously, the inspector's duty is to enforce the law but, like all public officials, he has a necessary discretion in the way he enforces it, subject to general departmental policy, public or judicial criticism and ultimately criticism in Parliament through the Secretary of State.

Wide as inspectors' powers are, it goes without saying that it is their strict duty not to exceed them, but to act strictly in conformity with the statutory procedures that are applicable (for example with any regulations issued under s20(3) about the taking of samples).

An important duty imposed upon the inspectorate (and also upon the Commission and Executive) concerns the handling of information obtained in the course of carrying out functions under the Act (known as 'relevant information').

By s28 a positive duty is imposed not to disclose relevant information without consent except for carefully defined official purposes. Breach of this duty, by inspectors or others, may be punished by a term of imprisonment not exceeding two years, or a fine or both under s33(4)(e). The exception being that there is a positive duty to disclose information to persons employed at the premises or their representatives in accordance with s28(8). This latter requirement is discussed in chapter 9.

Likely causes of public complaint against an inspector may either be for failure to perform his duties (made by the courts, employees or public)[13], for excessive zeal or for unauthorized action in performing his duties.

Since employees or other persons aggrieved by an inspector's failure to enforce the law cannot instigate proceedings themselves against an offender, their legal rights, if any, are only against the inspector and are limited in the extreme. An employee who claims that he was injured by the negligent performance of an inspector's powers may perhaps have a right against the inspector[14], and also against the authority that employs the inspector, (as an extension of the vicarious liability of an employer for the wrongdoings of an employee in the course of his employment) but there is no precedent for a similar remedy against an inspector who has altogether failed to enforce the law.

On the other hand a person such as an employer, who has directly suffered injury through the inspector's actions in exercise of his powers, may have a remedy open to him if the inspector acted either outside his authority or wrongfully in some way.

Section 26 acknowledges a right of action against an inspector and goes on to provide that the enforcing authority may indemnify the inspector, even in circumstances where he was acting beyond his powers, in cases where he honestly believed that his actions were within his powers and that he was acting in discharge of his duties. This extended right of indemnity is of obvious importance to employers, as the existence of a civil right of action would be pointless unless the defendant were in a position to satisfy a substantial award of damages. The widely extended powers of the inspectors under the Act, to seize and destroy articles, for example, make this provision necessary, even though it is unlikely that there will be many instances of wrongful or unauthorized activity on the part of inspectors. For the same reasons the fear of liability for wrongful actions is a matter which will obviously deter the inspectorate from lightly exercising the power to issue prohibition notices.

Private enforcement of the Act

Section 38 makes it clear that no employee, trade union official, safety representative, or interested private person may directly set the law in motion to secure compliance with the provisions of the Act, health and safety regulations or the earlier statutory provisions. However, a person who has suffered personal injury directly through breach of the law may still, in certain circumstances, have a 'cause of action'. Under s47, an injured person retains his right to bring a civil action for damages upon breach of any of the earlier statutory provisions. An injured employee has no right of action for breach of the general duties under the Act but may have a right under health and safety regulations, except in so far as the particular regulations provide otherwise, providing always that he is a person to whom a duty is owed under the provisions in question.

Where no action for breach of statutory duty arises, there may still be independent civil rights. In particular, the common law of negligence may be invoked in industrial accident cases, either as an alternative to a claim for breach of statutory duty, or in cases where there is either no statutory duty applicable, or a duty which does not give rise to civil remedies.

Civil actions apart, employees and their representatives cannot take part directly in the legal enforcement of the Act. The extent to which they may do so by means of cooperation and consultation with both employers and inspectors will depend on the way the Act's policy of worker involvement (discussed in chapter 9) develops in practice.

8 Health and safety and the personnel manager

A new role for the personnel manager

The demand which the Act makes both for trained and careful employees, and for well organized and safe systems of work implies that the personnel manager will be involved in health and safety to a greater extent than formerly. The basic philosophy of the Act, that all persons within an organization should be involved in the effort to achieve safer and healthier working conditions, cannot fail to involve the personnel manager. The Robens precept that safety ought to be regarded as a normal management function, and part of the responsibility of management at all levels, applies to the personnel manager as much as to any line manager. In particular, the responsibility under the Act for selecting and training employees to achieve and maintain the demanded levels of safety competence usually falls upon personnel managers (or the training officers).

It is likely that these changes in the law will lead to an increase in the number of organizations in which the personnel manager is appointed as the member of the senior management team upon whom the primary responsibility for safety, including coordination of the work of safety specialists, is devolved.

Personnel managers will therefore need to examine their safety responsibilities in the light of their other functions, particularly the drafting and implementation of contracts of employment, the training of personnel at all levels, and disciplinary and dismissal procedures.

The contract of employment

The relationship between the employer and his employees and their mutual rights and duties is, in matters concerned with safety (as in all others) dependent first upon the express and implied terms of the contract of employment. In the absence of contractual duties, the employer and employee, whatever their respective obligations under labour legislation, would have no private rights, either of enforcement or other redress against each other. The introduction of the Health and Safety at Work etc Act therefore required the employer to consider whether his employees' contracts of employment were generally satisfactory. A satisfactory contract of employment is likely to be couched in terms which adequately provide for the introduction, maintenance and enforcement of safe systems of work which the Act requires of employers in the discharge of their statutory duties.

It is conceded that safety is only one among other issues that fall within the contract of employment, but a firm that has not provided clear terms and conditions of employment, so that each employee knows the scope of his responsibility and his authority, is unlikely to be either efficient or safe. The advent of the Act was therefore an opportunity for reconsidering existing systems as much as for introducing new ones. The organization which attempts to treat safety in isolation, and address its employees specifically on this issue, is likely to run up against resistance as individuals suspect that they are being required to undertake new responsibilities and to sacrifice existing (particularly compensation) rights. It is, in fact, unlikely that many employees have had their contractual rights and duties materially affected by the Act, except in cases where individuals have been given totally different jobs, for example a foreman appointed as a safety officer. It should not have been necessary in most cases to negotiate new contractual terms in order to comply with the Act, provided that the employer's existing contracts were clear and accurate.

A feature of the Act is that each employee, from the most senior manager downward, is required to obey reasonable orders in the context of safety, and also to exercise personal competence in safety as well as in vocational skills. Regard should therefore be paid to the provision of clear job specifications and to the utilization of disciplinary procedures for those who do not comply with the criteria expected of them by the terms of their contracts.

Safety rules, the safety policy and the contract of employment

Many firms may well already have, or may wish to introduce, safety rules within their organization. These may be published in the arrangements for the implementation of a safety policy required under s2, or within a firm's handbook, or quite separately as codes of practice related to special tasks. However, where such rules are actually spelt out in detail it is advisable to refer to them and the place in the safety policy at which they may be found. There can be few firms who do not have some sort of safety rules, if it is only 'no smoking' signs. It is unlikely that these rules form contractual terms for individual workers, but they are likely to be deemed reasonable orders for the employer to give the employee, so that it would be a breach of the employee's contract to disobey them, and there is judicial support for this view[1]. The Court of Appeal was asked to decide whether railway workers engaged in an industrial dispute were in breach of their contracts of employment when they had 'worked to rule' in a way that put an unreasonable interpretation upon their employer's rule book. Their Lordships were of the opinion that the interpretation which the employees placed upon the provisions in their employer's rule book was not valid. They also held that, although the rule book was not itself part of the contract of employment of each individual employee, failure to obey it was a breach of contract. The words of L J Roskill summarize the position well:

> It was not suggested that strictly speaking this formed part of the contract of employment as such. But every employer is entitled within the terms and the scope of the relevant contract of employment to give instructions to his employees and every employee is correspondingly bound to accept instruction properly and lawfully so given. The rule book seems to me to constitute instructions given by the employer to the employee in accordance with that general legal right.

A firm wishing to raise its standards of health and safety will be well advised to review its rules on safety, to amend and supplement them when necessary, to make sure that they are known and understood by employees and finally to take care that they are actually observed in practice.

Relationship between safety rules and disciplinary rules

The Employment Protection (Consolidation) Act 1978 s1(4)(a) requires that the written particulars of terms of employment, which all full time employees are entitled to receive by the thirteenth week of employment, shall include a note specifying any disciplinary rules applicable to the employee or referring to a document which is reasonably accessible to the employee and which specifies such rules. Similarly, the employee is entitled to be informed as to the grievance procedure which he may invoke. However s1(5) of that Act expressly states that these requirements shall not apply to rules, disciplinary decisions, grievances or procedures relating to health and safety at work.

These provisions are difficult to understand in that no-one can doubt that it is desirable to spell out safety rules and to enforce their observance; moreover it is often difficult in practice to distinguish issues of safety from misconduct generally.

It is suggested therefore, that s1(5) is intended for the protection of the rights of the employee. The creation of disciplinary and grievance procedures often contain agreements for the maintenance of the 'status quo'. Thus without this subsection the situation might arise that an employee who reported an allegedly unsafe situation might be required to go on working with it until an enquiry discovered whether the allegation was correct.

The safety policy and the personnel manager

Section 2(3) of the Act requires every employer to set out in a written statement his general policy on health and safety at work, together with the organization and arrangements for carrying out that policy.

Many organizations, seeking to comply with the requirements of the law by the statutory deadline, issued policy statements which were incomplete and gave information concerning only the skeleton of their safety organizations. For example in large groups of companies, or in companies which operated from a number of sites, it was not uncommon to find that the statement related only to group policy and went no further than saying something like 'it shall be the responsibility of the managing director of each subsidiary to implement the company's policy within his area.' Sometimes the responsibility for implementing the policy was imposed upon a site manager. In such cases it is doubtful whether the company had fulfilled its statutory obligations under s2(3), until the named manager had in his turn produced a statement of his local organization and arrangements. Quite apart from any legal requirements, no company can be said to have achieved a safe system until this local task is completed. This is a task which appears to fall within the province of the site personnel office, to be carried out within the requirements of existing managerial structure at local level and with regard to the existing job specifications of individual employees.

The safety policy should be kept constantly under review. Not only should steps be taken to ensure that a copy of it is received and understood by each new employee, but the contents of the policy should be re-assessed from time to time to ensure that the policy accurately records the situation currently existing. Consideration should also be given to whether better arrangements could be made, even if the policy accurately records what is happening. It is undesirable to make alterations in the policy without consultation with employees particularly if the organization has safety representatives.

Consultation and disclosure

The Act places great emphasis upon the importance of disclosure of information, and consultation with employees. There are detailed requirements relating to consultation with employee safety representatives: this matter is dealt with in chapter 9. One question which the personnel manager ought to be asking is whether management has made provision for trained persons to represent them in the consultation process, as employee safety representatives are likely to be well informed as a result of the specialist training provided by their trade unions.

Training

The Act now expressly requires employers, as one of the matters to which they must attend in order to discharge their general duty, to provide instruction and training to ensure the safety of their employees. This requirement makes it desirable to review all aspects of the firm's existing training schemes and also to consider whether new training programmes need to be introduced to deal with other areas which have hitherto been neglected. In particular, attention should be given to the following.

(a) Induction courses

New employees, or employees starting work in a new capacity, need to be trained in the systems and practices of their firm. At the induction stage they should be made conversant with any safety rules which are of general application, in particular with emergency procedures for contingencies such as fire. They should also have pointed out the need for following such rules and the consequences of not doing so. Rules are more likely to be obeyed if the reason for them is understood, especially when training young employees.

(b) Training for the job

Training in the operation of machinery and equipment should be given not only when an employee starts work, but whenever new methods are introduced. Such training should include safety, by making the operative familiar with both the hazards and the means to protect himself against accidents and ill-health. Consideration should also be given to post-accident procedures, particularly when team work is involved; for example, there may be circumstances in which some employees may be placed in the position of having to assist an injured colleague. This is in addition to the need for first-aid certificate training for some employees[2];

indeed, the training required generally will include the procedure for obtaining the assistance of the trained first-aider.

(c) Specialist training for safety

The firm may well decide that it should appoint a person (such as a 'safety officer') with special responsibilities for safety, if it has not already done so, and to keep under review the status of the appointment. In particular, regard should be paid to the rest of the management structure and the relationship of the safety officer within this structure. The question may well be asked whether it is likely to be more effective to have one professionally trained safety manager at a fairly senior level or a number of safety officers, who have had some safety training, at foreman level, or a combination of the two. It may be necessary to ask, whether these should be full time appointments or whether these tasks should be carried out in conjunction with other duties. If other new appointments of this nature are being made, or if there are existing staff in post who have had some safety training, have the people concerned been properly trained for the task?

(d) Training in the requirements of legislation

The requirements of s2(2)(c), that the employer provide information and training, is wide enough to be interpreted as requiring the employer to train the employees in the understanding and observation of the statutory duties imposed upon them.

Before the Act, the law required[3] an employer to compensate an employee who had suffered injury because he had not understood or complied with the full extent of his duties under the Construction Regulations. The basis of liability was that the employer had failed to instruct the employee in the meaning of the regulations and his obligations under them. It seems unlikely that the Act requires a lower standard than this.

(e) Management training

The Act does not confine training requirements to the provision of training for operatives which enables them to look after their own safety; it also requires training of managers in order to ensure safety. Thus it takes into account the need to install and operate safe systems of work. Compliance with the Act therefore requires training of managers to enable them to introduce, operate and enforce safe systems of work.

(f) Refresher courses

Initial training may not be adequate if nothing is done thereafter to ensure that the matters dealt with during training are observed in practice, and that techniques are brought up to date in the light of changes in technology or the systems used in the organization. Therefore any effective training scheme must include refresher courses.

(g) Safety representatives

Clearly the introduction of employee safety representatives created a new training need. The TUC and some unions were active in planning and providing suitable training courses for safety representatives. An approved *Code of Practice for Time Off for Training of Safety Representatives* became operative on 1 October 1978. It provided that as soon as possible after their appointment, safety representatives should be permitted time off with pay to attend basic training facilities approved by the TUC or by the independent trade union or unions which appoint the safety representative. Further training, similarly approved, should be undertaken where the safety representative has special responsibilities or where such training is necessary to meet changes in circumstances or relevant legislation.

The Code of Practice did not attempt to lay down standards for the length of training but said that basic training should take into account the functions of the safety representative and should provide an understanding of the role of safety representatives, safety committees, and of trade union policies and practices in relation to:

(a) the legal requirements relating to health and safety of persons at work, particularly the group and class of persons they directly represent
(b) the nature and extent of workplace hazards and the measures necessary to eliminate or minimize them

(c) the health and safety policy of employers and organizations, and arrangements for fulfilling these policies.

Additionally the Code suggested that safety representatives would need to acquire new skills in order to carry out their functions, including safety inspections, and to use basic sources of legal and official information provided by, or through, the employer on health and safety matters.

The Code stated that when the trade union wants a safety representative to attend a course it should inform management of the course it has approved and supply a copy of the syllabus, indicating its contents, if the employer asks for it. It should normally give at least a few weeks' notice of the safety representatives it has nominated for attendance. The number of safety representatives attending training courses at any one time should be that which is reasonable, bearing in mind the availability of relevant courses and the operational requirements of the employer.

Unions and management are advised to reach agreement on appropriate numbers and arrangements, and refer any problems which may arise to the relevant agreed procedures. Sometimes the employer himself may provide the appropriate training[4].

It will be noted that the Code places the primary responsibility for defining the type of training needed for safety representatives upon the trade union movement; nevertheless it is suggested that the personnel manager should consider what, if any, further training, possibly related to the hazards and systems of the organization, should be provided for safety representatives beyond the basic statutory training.

Discipline

Operating a safety policy requires not only the introduction of a safe system of work but also its maintenance.

There can be little doubt that the best mechanism for encouraging the adoption of safe systems of work by employees is by training and by example. Individual workers will usually follow the pattern of behaviour set by their colleagues and superiors, but even in the best run organizations there tend to be the occasional individuals who prefer to persist in their own unsafe practices rather than follow the good example of others. In these presumably rare cases the employer needs to be able to take disciplinary action to bring this wrongful behaviour to an end.

The fact that an employee may be in breach of a duty imposed upon him by the Act, and be criminally liable[5], is of small comfort to an employer whose wish is not so much to punish the employee as to ensure the maintenance of a safe working environment, both for the wrongdoer and for other workers. At the very least, the employee who endangers his own safety is an annoyance, both for the example he sets and because he may well cause an accident which will involve the employer in expense and disrupt his business operations.

It seems clear that disregard of safe systems and practices, established to protect the safety of the employee and fellow employees, would amount to misconduct at common law and probably be regarded as a breach of an express or implied term of the employment contract. It is also clear that, in order to establish this point, an employer would need to be able to show that the employee knew of the requirement with which he had failed to comply. If the employer has laid down, and consistently sought to enforce, an organization and rules of conduct for the promotion of safety, it should not be difficult to establish that the employee was aware of his responsibilities.

Nevertheless, except in the most flagrant and serious cases of disregard for safety, the employer should follow the procedure recommended by the disciplinary code of practice for dealing with cases of misconduct; namely, to give the employee an initial warning (preferably in writing) setting out the circumstances. Only if the misconduct is repeated should the employer go on to consider the termination of the employee's contract of employment[6].

The right to suspend from employment for disciplinary offences may be made an express term of the contract of employment. If this is not done, no right exists at common law to suspend without pay for any reason[7]. It is not entirely certain whether an employer has the right to suspend a worker, even with pay, in the absence of an express contractual term[8].

Dismissal for unsafe conduct

In a case of substantial misconduct related to safety, an employer is entitled to dismiss the employee concerned. The employer should be able to justify the dismissal if challenged to do so in a proceeding for unfair dismissal before an industrial tribunal, provided that there was no evidence that the employer had condoned the behaviour of that employee, or of other employees in like circumstances, and that adequate warning had been given where necessary. It should be possible to show that such a dismissal was for one of the relevant permitted reasons falling under s57(2) of the Employment Protection (Consolidation) Act 1978; namely, that it was either for a reason which (a) related to the capability of the employee in performing work of the kind which he was employed to do, or (b) related to the conduct of the employee.

It must be stressed that, in accordance with the rules relating to unfair dismissal, it would be necessary to prove that a breach of safety rules was the reason for dismissal. Moreover, a tribunal would have to be satisfied that the employer acted reasonably in treating the matter as a sufficient reason for dismissing the employee. (Employment Protection (Consolidation) Act 1978, s57(3) as amended by the Employment Act 1980.)

As the law relating to dismissals stands at present, an employer cannot be compelled to retain in his employment an employee whom he considers unsafe; however, if an industrial tribunal takes the view that an employer had no reasonable grounds for dismissing an employee, dismissal is likely to be expensive and unlikely to improve industrial relations. For example, if an employer allowed 99 employees to smoke contrary to a rule, dismissal of the 100th would not be fair. Similarly, stories are told of foremen who almost daily warn particular men about failure to wear protective clothing and the warning becomes a matter of routine which is not taken seriously. In such circumstances, a sudden decision to enforce proper behaviour and to take the harsh disciplinary measure of dismissal is unlikely to be deemed fair. Nor would it be regarded as an adequate enforcement of a safe system of work in compliance with s2(2)(a) and (c).

It is possible that the provisions for worker involvement in safety matters through the appointment of safety representatives, will help to inculcate in employees generally the need to observe safety requirements and that the group may in turn exercise influence over the recalcitrant few.

Dismissal on grounds of ill-health

There may be occasions upon which a dismissal is fair because the employee's health is such that the employer cannot provide him with work without exposing him, and other workers or the public, to unwarranted risk of personal injury. For example, a car hire firm might well be able to justify the dismissal of a driver found to be liable to a diabetic coma while driving. The attitude of the industrial tribunals is not entirely certain on dismissals for health reasons. There are instances in which such a dismissal has been deemed unfair because there was suitable alternative employment available for the employee. Thus in the example given the tribunal may only consider the dismissal fair if it were not possible to offer the employee suitable employment in, say, the vehicle maintenance department or the reception office. The view emerging seems to be that before dismissing for reasons of ill-health the employer should endeavour to obtain a medical opinion and to discuss with the employee the needs of both parties. The dismissal may be fair, however, if it is in the interests of the employer's business[9]. It will also be fair if the ill-health of the employee is causing the employer to break his duty to provide reasonably safe working conditions for other employees[10].

There are instances where statute regulates the further employment of persons who have been exposed to health risks, in the employment where the hazard exists[11]. In cases of this kind, at common law, the employer might at best have suspended the employee or, at worst, dismissed him. Under the Employment Protection (Consolidation) Act 1978 workers who are suspended or dismissed from employment for medical reasons in compliance with certain statutory requirements, (including the occupational health requirements set out in Schedule I of the 1978 Act, and codes of practice made under the Health and Safety at Work Act), may be treated as having been unfairly dismissed, or will be entitled to remuneration as the case may be. It is the employer's duty under these provisions to find the employees alternative employment or to suspend them on full pay for up to 26 weeks. These provisions apply only while the employee is medically 'fit', apart from his occupational disability in that he is unable by law to follow his normal employment; should he fall ill, then he may be treated by the employer in the same manner as he treats any other case of employee sickness.

Incidentally, the fear of suspension and dismissal may well deter employees from participating in employers' schemes for medical inspection and testing. It might be advisable for employers to consider whether it would be beneficial to negotiate contractual terms under which an employee agrees to have regular medical inspections and to accept the decision of the company's doctor on his fitness for work in exchange for insurance rights to cover sickness or possibly early retirement.

Dismissal as a result of a prohibition notice

An employer may feel obliged to dismiss employees because a prohibition notice has been served, which prevents the use, until further notice, of the plant or premises at which his employees were engaged. Should this happen, and the continuation of the employer's business is not possible, it would appear that under the Employment Protection (Consolidation) Act 1978 s57(2)(d) the dismissal might be fair, since the employee could not continue to work in the position which he held without contravention on the part of his employer of a duty or restriction imposed by or under an enactment, provided that it is not possible to offer the employee alternative employment.

Alternatively it would seem that the situation might amount to a redundancy under s81 of the Employment Protection (Consolidation) Act 1978, which provides that an employee is redundant if his dismissal is attributable wholly or mainly to:

(a) the fact that his employer has ceased, or intends to cease, to carry on business for the purposes of which the employee was employed by him, or has ceased or intends to cease, to carry on that business in the place where the employee was so employed or
(b) the fact that the requirements of that business for employees to carry out work of a particular kind, or for employees to carry out work of a particular kind in the place where he was so employed, has ceased or diminished or are expected to cease or diminish.

It is hard to imagine that a dismissal as a result of a prohibition notice would not qualify under one or other of these heads if the employer could not offer suitable alternative employment, and if the employee in question was not disqualified from claiming on some such grounds as old age or lack of continuity of service.

It would, however, be in the employer's interests to try to avoid the interpretation that the situation was a redundancy for not only would he (in that situation) have to find the employer's contribution towards the redundancy payment, but he might, if he had a recognized union, have to consult with that union before dismissing the employees in question.

The employer may wish to consider the possibility of suspending his employees during the time the notice is in force, but this tactic is unlikely to be successful unless there is an express term in the contract permitting suspensions. In any event, an employer faced with a prohibition notice is likely to be caught, in the short term, by the guarantee payment provisions of the Employment Protection (Consolidation) Act, s12, even if he is contractually entitled to lay off his employees.

9 Health and safety and the worker

Employees and self-employed persons

The Act differs from previous safety legislation in that it protects almost the whole of the workforce. It does not distinguish for this purpose between employees and self-employed persons, nor between management and shop-floor, 'blue collar' or 'white collar' status. This comprehensive protection comes mainly from the imposition of obligations upon employers and controllers of premises, but the individual worker is also granted rights and is subject to responsibilities under the Act. Some provisions specifically create rights and duties for the self-employed (*see* s3) or lay upon an employer duties in respect of the health and safety of workers who are not his employees (*see* ss3 and 4): regulations and codes of practice made to implement the Act may further define and increase these duties. The major part of the labour force is likely to be working under contracts of employment, and so it is not surprising to find that a large part of the Act is concerned with the rights and duties of the employer and employee working under a contract of employment.

(a) Who is an employee?

This Act (like other legislation relating to employment) does not attempt any more enlightening definition of 'employee' than to say that he is an individual who works under a contract of employment or apprenticeship, whether express or implied and, if express, whether oral or in writing. Likewise, a 'self-employed person' for the purposes of the Act is an individual who works for gain or reward otherwise than under a contract of employment, whether or not he himself employs others.

There is nothing in the Act to give guidance on the method of distinguishing the worker who is an employee from the worker of independent status: both may well be working under contracts of broadly similar terms. The courts have acquired a great deal of experience in identifying the master–servant relationship and have produced a large amount of case law on the subject. Much of this case law is likely to be useful in helping to determine, in marginal cases, the nature of the employment relationship for health and safety purposes.

(b) Is the distinction between an employee and a self-employed person important?

It may well be that in many situations the distinction between an employee and a self-employed person is of little practical importance, (so far as the employer is concerned), since in either case his general duty will be to ensure, so far as is reasonably practicable, the health and safety of the worker (either under s2 or s3); and in any given case the nature and extent of his duty will depend upon the facts of the situation. It is not clear how far an employer will be required to instruct, train and supervise a self-employed person, or whether he would have any practical power to do so. However, in prescribed cases the employer will have to give information to those who are not his employees (whether they be self-employed or the employees of another) about such aspects of the way in which he conducts his undertaking as might affect their health and safety. In some instances he may be obliged to provide the same information in discharge of his general duties to (a) his own employees under s2 and (b) other workers under s3, even though there are as yet no particular regulations on the point in question[1]. There might be situations in which the employer could only discharge his duty of reasonable care to a self-employed person by giving him instructions similar to those which he must give to his own employees and to employees of others.

Similarly, the real distinction between the duty which the Act imposes upon the employee to take reasonable care for the health and safety of himself and other persons (s7), and the duty of every self-employed person to conduct his undertaking in such a way as to ensure, so far as is reasonably practicable, that he and other persons (not being his employees) who may be affected are not thereby exposed to risks to their health and safety, may not be immediately apparent in many factual situations. Admittedly the general purposes of the Act to promote occupational health and safety clearly require broadly similar general standards of careful behaviour from the employer in the provision of a safe system of work, irrespective of both his relationship with the worker and the nature of the worker's status: the worker, moreover, whether he be employed or independent, is also required to exercise reasonable care, personally, for himself and for others. The degree of initiative required will clearly vary according to the freedom of action of the worker in question, acting either as a member of a team of employees, or as a contractor subject only to control through the terms of his contract.

The manager as employee

Previous safety legislation has had limited objectives – generally the protection of manual workers in particular occupations. Protection of workers in non-manual or managerial occupations was largely fortuitous. It is important to bear in mind that the Act applies to all workers in all occupations and that persons occupying managerial positions are no longer excluded from the benefits, protection and obligations of the Act by the nature of their employment. Under this Act, every manager, like every shop-floor operative, is entitled to a safe system of work, and must be provided with such information, instruction, training and supervision as is necessary for his health and safety. As the relevant s2(2)(c) is concerned with ensuring the health and safety of employees generally, the manager is clearly entitled to such information, instruction, training and supervision as is necessary, both to protect him personally against the hazards of his employment and to fit him for his place in the management chain, so that he can play his part in ensuring the health and safety of those in any way dependent upon him.

Since managers are employees, they, like blue collar workers, are obliged by s7 to take reasonable care for the health and safety of themselves and of other persons who may be affected by their acts or omissions. They may also be placed under specific legal duties by regulations made under the Act; for example a supervisor might be named as the 'competent' person required by regulations to monitor a particular work situation within his area of responsibility.

Although the Act places emphasis upon management responsibilities for ensuring the health and safety of persons at work, it seems unlikely that managers (except at the highest levels) will be rendered personally liable for the failings of the organization which employs them: they will however be liable for their personal failures as employees in the performance of duties which have been specifically delegated to them, in fulfilment of their managerial responsibility. Thus they are most likely to be liable for failure in carrying out their duties under s7, interpreted in the context of their particular employment situation. There seems no reason to suppose that persons whose management position is below that of being an 'officer of the corporate body'[2], will be rendered personally liable for an organizational failure to identify hazards and implement safe systems for ensuring the health and safety of persons at work.

Where safe systems laid down by the company are not implemented due to the slackness of first line management, it is likely (quite apart from the personal liability of these employees, under s7) that the employer will still be liable for failure to exercise proper supervision and discipline to ensure that the systems existing on paper are implemented in practice.

Safety and the contract of employment

(a) The relationship between an employer's duties and the employee's duties

The relationship between an employer and his employees, and their mutual rights and duties is, in matters relating to safety as in all other matters, dependent on the express and implied terms of the contract of employment. In practice the duty imposed upon an employer to ensure the health and safety of employed persons and others will be discharged only by requiring each employee to take reasonable care, for his own safety and for the safety of others. In other words, the ability of an employer to perform his obligations under the Act may be jeopardized if the employee does not comply with the duties, both to act safely and to cooperate with others, imposed upon him by s7 of the Act. The knowledge that his employee is in breach of his statutory obligations is of little comfort to the employer who wishes to achieve safe working conditions. It is possible that an employee's default might provide the employer with a good 'defence' from liability under s36 of the Act, (which relates to offences due to the fault of another person). The Act, however, provides an employer with no direct means of enforcing criminal sanctions against defaulting employees, because only an inspector can institute a prosecution for an offence. Just as an employer cannot directly enforce s7 against an employee who fails to exercise reasonable care, or to cooperate, so the employee cannot legally enforce the 'rights' created in his favour by the duties imposed upon his employer under s2. He can only enforce these rights if they have become a part of his contract of employment.

(b) The employee's contract

The requirements of s2(2)(c) that an employer provide such information, instruction, training and supervision as is necessary to ensure the health and safety at work of his employees, may be complied with through the contract of employment itself. An employer could, for example, go some way towards discharging his obligations to inform the employee by incorporating into the contract an indication of the particular hazards to which the employee is likely to be exposed and the appropriate safeguards to minimize the risks. For example, the contract might describe the job in terms which imply that it is a condition that the employee holds himself out as having certain skills to deal with specific hazards, for example a qualified shotfirer, employed as such, would be expected to assimilate and observe the obligations imposed upon him personally by the Shotfiring Regulations[3].

(c) The rights of the employee

The major purpose of the Act is to protect employees and others by imposing duties for their protection on a variety of persons. In this broad sense the Act entitles an employee to expect that his employer will do all that is reasonably practicable to ensure his (the employee's) health and safety. He is also entitled to expect that other employees, self-employed persons, the occupiers of premises to which he may be sent to work (provided they are not domestic premises), the manufacturers and suppliers of goods and substances, will all take positive steps to try to ensure that he is not injured while going about his work. An employee is, incidentally, entitled to expect his fellow workers to have regard for his safety. The provisions of the general duties do not give the employee any legal rights, strictly speaking, since he is not able to institute any proceedings either at civil law (s47), in the event of receiving personal injury, or in a criminal court (s38), for his general protection. All an employee can do to enforce his rights, is either to seek the support of the Executive and rely upon the inspector appreciating the merits of his grievance, or involve his safety representatives[4].

Independently of any rights bestowed upon an employee by the Act, an employer is clearly under a contractual duty to provide reasonably safe working conditions. When these are not provided an employee may arguably regard himself as released from his duty to work under the contract. If the place of work or the system of work with which an employee was expected to comply were themselves unsafe, and the worker walked out, he might possibly lodge a complaint for 'constructive dismissal' and be regarded as having been 'unfairly dismissed' as defined in the Employment Protection (Consolidation) Act, 1978, s55(2)(c)[5].

Similarly, failure on the part of an employer to provide an employee with a safe system of work may provide the employee with the basis for a claim at common law. This is a right which the employee cannot readily enforce, unless he wishes to claim damages after suffering an injury as a result of the breach of duty. Apart from this, he can only treat the contract as repudiated, and withdraw his labour if he does not receive safe working conditions. It is interesting to speculate whether an employee who acted in this way could nevertheless obtain damages from his employer for breach of contract. Certainly he would not be able to bring an action for enforcement of the contract with safe working conditions. He would be more likely to use his employer's breach as a defence if it were claimed by the employer that the employee was himself in breach by withdrawing his labour[6]. In this situation, the 'qualified' employee[7] would be wiser to seek reinstatement under the statutory scheme for unfair dismissal than in the common law courts.

At present there would appear to be no real protection against the dismissal or victimization of an employee who campaigns for safety at his place of work. It is true that, in the case of such a dismissal, the employer might well have difficulty in establishing that the dismissal was fair, unless he could satisfy a tribunal that the principal reason for dismissal was other than the safety activity. It is interesting to speculate on whether the unfair dismissals provisions would provide protection for employee safety representatives appointed by the union under s2(4) of the Act and the Safety Representatives and Safety Committees Regulations 1977; or for union members who display militancy in association with a union in respect of safety matters[8].

The justification for this argument is that such safety activity fell within 'taking part ... in activities of an independent trade union', which is protected activity under s58(1)(b) of the Employment Protection (Consolidation) Act 1978.

Where a worker claims that he has been discriminated against, but not dismissed, for safety activity, his statutory protection is equally weak. The Employment Protection (Consolidation) Act, s23(1)(b) provides protection for an employee who has action short of dismissal taken against him by his employer in order to prevent or deter him from taking part in the activities of an independent trade union at any appropriate time, or penalizing him for doing so. Once again, one may speculate whether these provisions would protect either an employee or a union appointed safety representative, since it is not clear that the safety representative performing his duties, or the employee for that matter, is taking part in the activities of an independent trade union.

It appears that the provision is a weak one, apart from the problem of proving the discrimination against the employee in question, since the only remedy lies in the power of a tribunal, under s24 of the Employment Protection (Consolidation) Act to declare the complaint well-founded and to award compensation. Unless the employee can establish something definite, like a failure to promote him, it is hard to see how his loss can be assessed for compensation purposes. The provision is also weak because it only protects the employee, and therefore does not assist the self-employed worker.

(d) The duties of the employee

The contract of employment is also the source of the duties that may be required of an employee. These contractual duties may include detailed requirements for the employee's own protection, and for the protection of others. Senior management staff are employees of the company and their contracts may impose quite substantial responsibilities upon them for the safety of other employees.

The general duty of an employee under the Act is to take reasonable care for his own safety, and for that of others, and to cooperate with his employer, or any other persons, so far as is necessary, to enable the others to perform the duties imposed upon them by safety legislation. It is likely that the terms of a contract of employment will throw considerable light upon what the individual employee can be expected to do to display reasonable care for the safety of others. Accordingly, an employee might well be prosecuted for behaviour which in another context would be regarded as a breach of contract of employment. It may also be that the statutory protection given to persons taking part in trade union activities will not be extended to employees who, in the course of a strike take steps, such as turning off a hospital boiler, which are not only in breach of contract, but also (in breach of the Act), cause danger to others:

> 'It would be monstrous to think that Parliament could have created a situation where any individual, trade unionist or otherwise, could place the health and safety of young people and old people at risk, and then come along and say you can do nothing to me, you cannot even dismiss me.' (per John Harvey QC before an industrial tribunal on an application for interim relief (Employment Protection (Consolidation) Act 1978, ss77–80), reported in *The Daily Telegraph* Saturday 23 December, 1978.)

It may well be that a senior manager is more vulnerable to prosecution under s7 than are more junior employees, owing to the wider scope of the responsibilities imposed upon senior management.

Worker participation

An important aspect of the Robens philosophy is worker involvement in the task of introducing and maintaining safe conditions at the workplace. The Act implemented this by requiring that workers (*a*) be kept fully informed on all matters relating to their health and safety, and (*b*) be consulted by employers about the organization of safety at the workplace.

(a) Information

An employee is entitled to expect that his employer will provide information necessary to ensure, as far as is reasonably practicable, health and safety at the workplace (s2(2)(c)). The subsection does not, it will be noted, stipulate to whom this information shall be provided. It places the duty in general terms, leaving some doubt as to whether the information should be provided to each employee personally, to the workers' representative or to another person, such as the supervisor, the safety representative or the company doctor, so long as the provision is effective. Would an employer for example, be discharging his duty to inform, if he merely informed the company doctor that his workers were exposed to a possible risk of cancer, and instructed the doctor to monitor the health of those workers who were at risk? Or would the proper discharge of this part of the employer's duty require that he also inform the employees themselves of the steps he was taking to monitor their health and his reasons for doing so? In this particular instance the employer could not fully perform his duty while keeping his employees in ignorance, since their cooperation in identifying the symptoms of illness would be essential. Furthermore, an uninformed worker, who changed employment in this situation, might be put to unnecessary risk because of his ignorance. Nevertheless, there may be occasions when the safety of employees could be ensured, within the meaning of s2(2)(c), by means of information given to a manager but withheld from the individual employees.

Such a decision by an employer to withhold information from employees might be lawful, if doing so did not expose them to extra risk; but it is questionable whether, in the long run, it is in accordance with good industrial relations policy or is in the spirit of this Act, which is clearly intended to foster maximum worker involvement in matters of health and safety at work.

Moreover, there is another important reason why employers should not withhold relevant information from individual workers and their safety representatives; namely, that the inspector is bound by s28(8) to give safety information to persons employed or their representatives. This information comprises both factual information relating to the work premises and information on any action the inspector has taken, or proposes to take, concerning the premises. Clearly, in the interests of good industrial relations, an employer should try to ensure that his employees learn about work hazards from him rather than from an inspector.

(b) Consultation defined

The Act gives no definition of consultation, although it places considerable emphasis upon the use of consultation as a means of worker involvement in accident prevention. The original Industrial Relations Code of Practice published in 1971 may well have both influenced and reflected contemporary views on industrial relations. It described 'consultation' as 'jointly examining and discussing problems of concern to both management and employee. It involves seeking mutually acceptable solutions through a genuine exchange of views and information' (para 65 *et seq*).

The code rightly drew a distinction between consultation and negotiation although it recognized that they were closely related processes. The distinction, in the context of this Act, is that safety remains management's responsibility; consequently management cannot discharge this responsibility by adopting unsatisfactory safety standards merely because they are acceptable to the employees. Management must assess the information and advice received through the consultative process and decide whether safety standards based upon it would amount to the proper discharge of management responsibilities under the Act. This distinction, between consultation and negotiation, seems worthy of examination, because with the introduction of employee safety representatives with statutory functions the consultative process under the Act has been pushed much nearer to the main stream of industrial relations than Robens and the originators of the legislation may have envisaged. There may be a danger of forgetting that in the parent legislation the powers of safety representatives to influence management decisions were seen as relatively limited ones.

(c) Distinction between 'employee consultative' and 'management advisory' committees

Employers wishing to keep problems of health and safety under review may decide to set up a committee charged with the duty of identifying particular employment risks and advising management on how to control them. Such a body may quite appropriately, for some purposes, be named a safety committee or consultative committee. But because it merely advises management in safety matters, it should not be confused with the statutory safety committees which must be set up if required by safety representatives appointed in accordance with the regulations.

10 Employee safety representatives

The Safety Representatives and Safety Committees Regulations were laid before Parliament in March 1977 and came into force on 1st October of the following year. They make provision for safety representatives to be appointed in accordance with s2(4) of the Act and they set out the cases in which recognized trade unions may appoint safety representatives from among employees and the functions of those safety representatives when appointed.

Much of the factual information in this chapter is drawn directly from these regulations so for convenience the regulations have been reproduced in full in appendix I(c) commencing page 149.

The Health and Safety Commission has also approved a code of practice on the rights and duties of safety representatives and this also came into operation on 1 October 1978. In addition the Commission issued two sets of guidance notes, the first offers advice to all concerned with the appointment and functioning of safety representatives, and the second to give advice concerning the setting up and functioning of safety committees.

All these documents have been brought together in a single booklet, entitled *Safety Representatives and Safety Committees*, which is available from HMSO.

The status of the regulations

The regulations and supporting code of practice may be regarded as a legal framework within which employers and trade unions may negotiate a system of safety representatives and safety committees suitable to their particular needs. The intention is not to restrict unnecessarily the freedom of employers and trade unions to make their own arrangements; therefore nothing in the regulations or the code is intended to prevent employers and employees continuing existing agreed arrangements which remain satisfactory to both sides, or from drawing up, on a purely voluntary basis, working arrangements which are quite distinct from the scheme of the regulations.

In other words the regulations, although not mandatory, are similar to other statutory 'floors of rights' and like them set out the minimum to which employers covered by the regulations may be required to conform. In this instance it is not obligatory for an employer to adopt the scheme set out in the regulations, unless the employees acting through their trade unions wish it; while employees may request their employer to implement an alternative scheme which they find preferable, they have no legally enforceable power to demand any system other than that set out in the regulations: nor will the inspectorate require the regulations to be implemented where an organization has found a system which is acceptable to its recognized trade unions.

Thus the regulations leave employees free to adopt a number of strategies on safety representation. They may:

(a) take no action
(b) oblige the employer to provide the legal minima set out in the regulations
(c) obtain the legal minima as in (b) and, additionally, seek to negotiate some further rights (for example increasing the frequency of inspections, see below)
(d) seek to negotiate some quite separate and distinct system of safety representation.

Appointment and jurisdiction of safety representatives

The regulations provide that a recognized trade union may appoint safety representatives from among employees in all cases where one or more employees are employed by an employer; except that this right is not accorded in the case of persons employed in a mine. The persons appointed need not be employees of the employer concerned when they represent members of the British Actors' Equity Association or of the Musicians' Union, owing to the peripatetic nature of employment in these occupations. Miners were excluded from the ambit of the regulations because they had had, for many years, a separate system affording comparable rights, under the Mines and Quarries Act 1954[1].

A 'recognized trade union' means an independent trade union as defined in s30(1) of the Trade Union and Labour Relations Act 1974, which the employer concerned recognizes for the purpose of negotiation relating to, or connected with, one or more of the matters specified in s29(1) of that Act in relation to persons employed by him, or in cases which ACAS had made a recommendation for recognition under the Employment Protection Act 1975, before this system of recognition was repealed by the Employment Act 1980 s19. It is not outside the bounds of possibility, although perhaps unlikely, that a hitherto unrecognized trade union might seek recognition expressly for the purpose of appointing safety representatives.

The appointment of an employee as a safety representative becomes effective when a trade union notifies the employer in writing that the particular appointment has been made. A person so appointed shall, so far as is reasonably practicable, either have been employed by his employer throughout the preceding two years, or have had at least two years experience in similar employment.

Employers have no power to influence the appointment, or non-appointment, of particular employees as safety representatives for the workplaces under their control. They can, however, negotiate with the trade unions concerned on the number of safety representatives to be appointed or on the 'constituency' (the area and employees for which he will be responsible) of any safety representative who is appointed. Likewise, it is for the trade unions to decide which of the unions recognized by a particular employer shall have their own representatives at his workplace, or whether low membership indicates the expediency of one representative being taken to represent the members of several unions. In practice safety representatives, once appointed, may well claim to represent, or at least speak for, those non-union employees in the workplace which are subject to the same risks as their constituents.

The guidance notes suggest that when consideration is being given to the number of safety representatives to be appointed in a particular case, appropriate criteria would include:

(a) the total numbers employed
(b) the variety of different occupations

(c) the size of the workforce and variety of workplace situations
(d) the operation of shift systems
(e) the type of work activity and the degree and character of the dangers inherent in the operation.

A safety representative is appointed in respect of a 'workplace' which means any place or places where a group or groups of employees he is appointed to represent are likely to work, or which they are likely to frequent in the course of their employment, or incidentally to it. A 'workplace' may comprise part or all of an employer's premises. Difficult decisions may have to be taken in large establishments, or establishments where a number of distinct processes are involved, as to whether a particular safety representative's 'constituency' is co-extensive with the employer's premises, or is confined to certain areas or workshops; these are matters for negotiation, particularly if there are rival unions involved. Where the 'workplace' is not under the employer's control (for example electrical contractors employed in re-wiring the premises of another employer) the statutory functions of safety representatives of both employers will be limited, as representatives have no legally enforceable rights against any person other than their own employer.

An employee ceases to be a safety representative when:

(a) the trade union which appointed him notifies the employer in writing that his appointment has been terminated
(b) he ceases to be employed at the workplace: if, however, he had been appointed to represent employees at more than one workplace he will not cease to be a safety representative so long as he continues to be employed at any one of them
(c) he resigns as safety representative.

Removal of safety representatives

An employer has no power to remove a safety representative, who, in the employer's opinion, fails to perform satisfactorily the functions for which he was appointed. Nor can an employer dismiss a safety representative who shows too much zeal in the performance of his duties, because of the stringent general law relating to unfair dismissal. As far as discipline and dismissal are concerned, a better view would seem to be that a safety representative is protected as much as, but no more than, any other employee with the same service with the employer[2].

Employers can, of course, dismiss a safety representative for some reason not connected with the exercise of his rights as a safety representative, such as misconduct or redundancy; and no doubt employers can apply disciplinary sanctions to employees who use the status of safety representative in a manner incompatible with the duties of the tasks they had contracted to perform.

It is significant that safety representatives incur no civil or criminal liability under the regulations through performing their statutory duties, but of course remain, like all other employees, subject to the general criminal provisions of the Act while at work. It would appear that safety representatives are given no extra rights or duties, over and above those bestowed contractually on employees generally to interfere with their employer's system of work to deal with a hazard, real or supposed, at the workplace; though the opposite view has occasionally been argued.

Functions of safety representatives

The primary function of safety representatives appointed in accordance with s2(4) of the Act is to represent employees in the consultations with employers provided for in s2(6). Section 2(6) requires every employer to consult safety representatives with a view to making and maintaining arrangements which will enable the employer and his employees to co-operate effectively in promoting and developing measures to ensure the health and safety[3] at work of the employees, and in checking the effectiveness of such measures. In addition to this general function every safety representative is empowered to:

(a) investigate potential hazards and dangerous occurrences at the workplace (whether or not they have been drawn to his attention by his constituents) and to investigate the causes of accidents at the workplace
(b) investigate complaints made by any of his constituents relating to that constituent's health, safety or welfare at work
(c) make representations to the employer on matters arising out of paragraphs (a) or (b) above
(d) make representations to the employer on general matters

affecting the health, safety or welfare at work of his 'constituents'

(e) carry out inspections in accordance with the regulations

(f) represent his 'constituents' in consultations at the workplace with inspectors from the Health and Safety Executive or from any other enforcement agency

(g) receive information from inspectors in accordance with s28(8) of the Act

(h) attend meetings of safety committees in his capacity of safety representative in connection with any of the above matters.

The code of practice advises safety representatives to bring to the employer's notice, normally in writing, any unsafe or unhealthy conditions, working practices or unsatisfactory arrangements for welfare at work which come to their attention as a result of formal inspections or through day to day observation. The making of such a report to an employer cannot be taken to imply that all other conditions or working practices are safe and healthy, or that the welfare arrangements are satisfactory in all other respects.

The code infers that the making of a written report to an employer by safety representatives does not preclude bringing the same matters to the attention of an employer or his representative by direct oral approach, particularly in situations where speedy remedial action is called for. The code also states that it will be appropriate for minor matters to be made the subject of direct oral discussion with employers or their representatives without the need for a prior written approach.

The establishment of a system through which safety representatives may communicate with their employer in order to perform their functions may well be a matter for negotiation between an employer and the trade unions. Particular care should be taken to identify those with whom safety representatives should first discuss safety problems, and also to settle the procedure which should be followed by the employer's designated representatives in responding to, and resolving, the safety problems identified by safety representatives. Clearly, it is desirable that the employer's 'safety respondents' should be persons who have received training both in hazard identification and in the art of successful communication with safety representatives. It is suggested that the supervisor for the area concerned is the person whose experience will qualify him for this task, rather than a safety or personnel officer, either of whom are likely to be too removed to be best equipped to deal efficiently with the hazard.

The regulations also bestow other rights upon safety representatives which are uncertain in scope, and for this reason have caused some concern to management. Rights of this description which call for close examination include:

(a) the right to inspect the 'workplace'
(b) the right to inspect 'documents'
(c) the right to receive information.

Inspection of the workplace

The regulations provide that safety representatives shall be entitled to inspect the workplace, or part of it, if they have given the employer or his representative reasonable notice of their intention to do so, and have not inspected it, or the part in question, in the previous three months. They may, of course, carry out more frequent inspections by agreement with the employer. Where there has been a substantial change in the conditions of work since the last inspection (whether because of the introduction of new machinery or otherwise) or where new information, relevant to the hazards of the workplace, has been published by the Health and Safety Executive since the last inspection, safety representatives – after consultation with the employer – shall be entitled to carry out a further inspection of the relevant portion of the workplace.

Where there has been a 'notifiable accident' or 'dangerous occurrence' (within the meaning given to these expressions in the Notification of Accidents and Dangerous Occurrences Regulations) in a workplace, or a notifiable disease has been contracted there, safety representatives may carry out an inspection of the relevant workplace, providing they leave conditions undisturbed, so as not to frustrate subsequent inspection by inspectors from the Health and Safety Executive.

So far as is necessary for the purpose of determining the causes of an accident, safety representatives may also inspect any other part of the workplace. However, they may only carry out an inspection if it is safe so to do, and if the interests of their constituents are involved in the inspection. Where it is reasonably practicable to do so, they are required to notify the employer or his representative of their intention to carry out these inspections. An employer may be present at inspections or investigations, but his presence is not essential under the regulations; thus an employer cannot frustrate an inspection or investigation by means of protracted negotiations over the timing or other details in relation to it. An employer is required to provide such facilities and assistance for carrying out inspections as the safety representatives may reasonably require, including facilities for independent investigation, and for private discussions with the employees concerned. The guidance notes give examples of pro formas which might be found suitable for recording the details of any inspections that have been carried out, and for notifying employers of any unsatisfactory matters which have been discovered in the course of an inspection.

Employers may consider it advisable to agree with unions' programmes for inspections which employer's representatives attend: these inspections could be short, constituting in-depth surveys of limited areas of the workplace or concentrating upon particular hazards. Almost certainly the firm's safety officer should either be present at such inspections or, at least, available both at the time and immediately afterwards in order to consider findings without delay. The total number of safety representatives entitled to take part in these inspections should also be agreed in advance.

Inspection of documents

Safety representatives who have given an employer reasonable notice are entitled to inspect, and take copies of, any document relevant to the condition of the workplace or to the safety of their consituents. This right applies to any document which the employer is required to keep by virtue of the relevant statutory provisions, always excepting a document consisting of, or relating to, the health records of an identifiable individual.

This provision is of fairly limited scope and relates only to the records which an employer is required to keep. It is not, however, necessarily the only documented information which an employer may be required to provide to safety representatives; the extent to which the representatives' general right to receive information from the employer entitles them to be given access to other records remains a matter of controversy, until settled by a court.

Provision of information to safety representatives

The regulations require employers to make available to safety representatives the information necessary to enable the representatives to fulfil their functions. The guidance notes suggest that, in discharge of this duty, employers will be obliged to furnish representatives with information over and above that which they are obliged to supply to employees generally under s2 of the Act.

The regulations state that employers need not provide information that falls into the following classes:

(a) information, the disclosure of which would be against the interests of national security
(b) information an employer could not disclose without contravening a prohibition imposed by, or under, an enactment
(c) information relating specifically to an individual
(d) information, the disclosure of which would, for reasons other than its affect on health, safety or welfare at work, cause substantial injury to an employer's undertaking; or, where the information was supplied to him by some other person, to the undertaking of that person
(e) information obtained by an employer for the purpose of bringing, prosecuting or defending any legal proceedings.

Although this list is not dissimilar to the exceptions contained in the Employment Protection Act 1975 to the legal requirements for the disclosure of information for the purposes of collective bargaining, it remains unclear in what circumstances information relating to health and safety may be withheld. For instance, is an employer required to disclose medical records of an employee who has given his consent to the disclosure? Perhaps the better view is that the employer is obliged to do so, unless the records in dispute are records required to be kept under statute. Again, there may be difficult questions concerning information withheld under (d) above. Safety representatives might, for example, request the formula of a commodity marketed by their employer, under the suspicion that it contained a substance injurious to the health of employees required to handle it in the course of their work. It is unclear whether or not the employer in question would be entitled to withhold this information on the ground that the formula was secret and of value to competitors. Lastly, the extent of the protection given under paragraph (e) above is uncertain, because much information obtained during accident investigations by management may be collected as a precaution against the possibility of litigation, however unlikely this may have seemed at the time; it is not clear whether the protection of the exemption could be invoked against disclosure if the demand for it were made at a time when litigation seemed only a remote possibility. The House of Lords has, in litigation, ordered the discovery to a widow of a report prepared by her late husband's employer as a matter of routine, after an enquiry concerning the circumstances of his death at work: their Lordships did not regard the report as entitled to the protection against discovery which is accorded to documents prepared solely for the purposes of litigation[4]. It has been suggested that this ruling effectively entitles safety representatives to see all such reports, but this may be too sweeping[5].

The code of practice suggests that information provided to safety representatives should include:

(a) information about the plans and performance of the undertaking and any changes proposed in so far as they affect the health and safety at work of employees
(b) information of a technical nature about hazards to health and safety, and precautions deemed necessary to eliminate or minimize them, respecting machinery, plant, equipment, processes, systems of work and substances in use at work, including any relevant information provided by consultants, designers or by the manufacturer, importer or supplier of any article or substance used, or proposed to be used, at work by their employees
(c) information which is kept relating to the occurrence of any accident, dangerous occurrence or notifiable industrial disease and any statistical records relating to such accidents, dangerous occurrences or cases of notifiable industrial diseases
(d) any other information specifically related to matters affecting the health and safety at work of employees, including the results of any measurements taken by the employer or persons acting on his behalf in the course of checking the effectiveness of his health and safety arrangements
(e) information on articles or substances which are issued to outworkers.

In matters on which routine information is available, it may well be convenient to enter into agreements with the safety representatives or their trade unions for the provision of relevant information on a regular basis; for example information about accident frequencies might be supplied on this basis.

Facilities

Safety representatives and their trade unions are likely to press employers to enter into agreements for the provision of facilities for the performance of their functions. For example, they may request the use of a telephone, duplicating and typing facilities, a library of statutory and other safety literature and, perhaps, a suitably furnished office. It may be, of course, that the facilities already granted to trade unions by a particular employer are adequate to accommodate the additional requirements of the safety representatives; if this is not the case the employer should agree to provide 'reasonable facilities'. What are reasonable is likely to vary with the size of the workplace, the nature of the work undertaken there, the severity of the safety problems encountered, and the number of employees at risk.

Time off for safety representatives

Employers must permit safety representatives to take time off (with pay) during their normal working hours, so far as proves necessary for the purpose of performing their functions, and to enable them to undergo such training in aspects of their statutory functions as may be reasonable in all the circumstances.

As the regulations are not specific on this right to time off, employers will find it essential to negotiate with trade unions the time allowed to safety representatives for the performance of their tasks. A rigid allocation of time will not be satisfactory in many instances, since investigation of accidents, and other emergencies, may well ensure that the amount of time the safety representative devotes to the performance of his statutory duties will vary from week to week. An employer might consider negotiating a global allocation of time for the work of all the representatives functioning in his establishment, and leave it to the unions to decide how many hours should be allocated to each individual representative. The provision of time off for safety representatives is now subject to a code of practice.

If the employee functioning as safety representative is not salaried, it will be necessary to set up a mechanism to ensure that he is paid for the time in which he performs his statutory functions. If he normally receives a variable income derived from piecework, he must be allowed his average hourly earnings.

A safety representative whose employer has not permitted him to take time off for the performance of his functions, or has failed to pay him for the time he has devoted to them, may complain to an industrial tribunal within three months of the date of the employer's failure. An industrial tribunal that finds a complaint of such failure to be well-founded may make a declaration to that effect, and also award compensation to be paid by the employer to the employee. The compensation shall be of such an amount as the tribunal considers just and equitable in all the circumstances, having regard to the employer's default in failing to permit the employee to take time off, and to any loss sustained by the employee which is attributable to the employer's conduct.

It has been suggested that the industrial tribunal might, because of its power to resolve questions relating to payment of safety representatives, in practice become the forum for resolving more fundamental issues relating to the system of workplace safety representation, since there is no other mechanism for resolving these disputes in the civil courts. According to this suggestion, an employee might complain that he had not been given time off with pay in order to resolve the issue of whether he were indeed a safety representative at all. Similarly, an employer who registered disapproval of what he deemed an excessive number of safety representatives at a workplace by making only minimal time allowances to individual safety representatives might have this policy challenged. While these are possibilities there is no evidence that this is in fact happening. In fact, so little use has been made of this procedure since 1978 that it is still too early to decide whether or not it will be employed to any extent to resolve such management–union conflicts over the recognition of particular safety representatives.

As has been noted elsewhere (*see* chapter 8) the trade union movement has made a considerable effort to provide training courses to ensure that safety representatives are properly prepared for their functions. This does not exonerate the employer from providing additional special training and the relationship between the employer's and the union's prerogatives in training matters has been considered by the Employment Appeal Tribunal[6].

Safety committees

The main thrust of the regulations is directed towards the appointment and functioning of safety representatives; so that the establishment, constitution and functions of safety committees receive scant attention in the regulations. Many organizations will find this particular statutory priority the reverse of the system to which they have been accustomed, since formerly the safety committee tended to be the focal point in an employer's strategy for worker involvement in workplace safety.

Under the regulations it is provided merely that an employer shall establish a safety committee when at least two safety representatives request him in writing so to do. The employer on receiving such a request must consult with the safety representatives who made the request and with the representatives of the recognized trade unions whose members work in any workplace which will come within the jurisdiction of the proposed committee. The committee must be established within three months of the employer's receipt of the request. It remains for the employer to decide the composition and function of the committee and the workplaces that will fall within its jurisdiction: having made these decisions, the employer must inform the employees concerned by posting a notice.

Safety committees in the past have not always functioned in a satisfactory and effective manner, and there is no doubt that, if statutory committees are to serve a useful purpose, management will have to give careful attention to their composition and functions. Safety committees should neither serve as an arena for industrial conflict, nor as places where urgent matters receive prolonged and inconclusive discussion. It is important, therefore, that the management representatives on safety committees should be senior officials with executive powers and personal accountability in respect of safety matters, and it is always necessary to ensure that the agenda of the meetings should receive careful planning. Ideally a safety committee should be so conducted that it serves as an organ for planning long term safety strategies, rather than a forum for matters of day to day routine; the latter are better disposed of between safety representatives and management representatives at either formal or informal meetings.

The guidance notes suggest that the functions of a safety committee might include:

(a) the study of accident and notifiable disease statistics and trends so that reports can be made to management on unsafe and unhealthy conditions and practices, together with recommendations for corrective action
(b) examination of safety audit reports on a similar basis
(c) consideration of reports and factual information provided by inspectors of the enforcing authorities appointed under the Health and Safety at Work Act
(d) consideration of reports which safety representatives may wish to submit
(e) assistance in the development of works safety rules and safe systems of work
(f) a watch on the effectiveness of the safety content of employee training
(g) a watch on the adequacy of safety and health communication and publicity at the workplace
(h) the provision of a link with the appropriate inspectorates of the enforcing authority.

Enforcement of the safety representatives and safety committees' regulations

The analysis of the nature and provisions of the regulations (particularly the evidence on the extent to which they are based upon and relate to, the existing industrial relations procedures) leads to the conclusion that the whole system of worker involvement envisaged in the regulations will be dependent upon and enforced through, the sanctions usually associated with the collective bargaining process. This conclusion is certainly true as far as the procedures for the appointment and functioning of the representatives are concerned. Nevertheless, it is possible that the Health and Safety Executive's inspectorate might be called in to help resolve a situation in which an allegedly recalcitrant employer was failing to grant safety representatives their statutory rights and facilities. In such a situation the inspector would be empowered, if he were satisfied that the regulations were not being observed, to serve an improvement notice upon, or even prosecute, the delinquent employer. However it is unlikely that these legal sanctions will be resorted to in many instances. It is far more likely that the conflict will be resolved by the usual procedures for the resolution of disputes.

Safety representatives are more likely to call in the inspectorate when a hazard has been identified in the course of their duties, and which the employer has refused to rectify. In such a situation the position would not be substantially different from that resulting from employees' complaints under earlier safety legislation. The major difference will be, not in the powers of the inspectorate to deal with the complaint, but in the enhanced position of the employees themselves. The likely expertise of safety representatives in identifying hazards should not be underestimated in view of the substantial training provision made for them by the trade union movement. This expertise makes it far more likely that safety representatives will be able to identify hazards and breaches of legislation, while their consultative rights and protection against dismissal make it likely that the final resolution of a safety dispute will in many instances be settled between the two sides, with reference to the inspectorate becoming a matter of last, rather than first, resort.

Non-unionized workplaces

It is inherent in all that has been said in this chapter that there is no statutory provision for safety representation in non-unionized workplaces. In establishments where there is no significant union presence the employer may make arrangements for selection of employee representatives and operate a scheme which may have some or most of the features of the statutory scheme. The Health and Safety Commission has issued a short guidance note expressing the hope that employers will do this. It is unlikely, however, that such a voluntary scheme could be operated in conjunction with, or parallel to, a statutory scheme in establishments where one or more groups of worker are represented by a recognized trade union any more than such a 'mixed system' can be employed in relation to other workplace issues like pay and hours of work.

11 Management responsibilities under the Act

Legal and practical responsibilities

A most important object of the Act is to ensure that the promotion of health and safety at work is regarded as an essential function of good management. The Act sought to achieve this objective by laying stress on safe systems of work and by the responsibilities it imposes upon employers and managers.

This chapter contains an analysis of the ways in which these responsibilities may be discharged within a corporation, and then goes on to examine the impact of these responsibilities upon managers as individuals. For this purpose it is useful to distinguish between the 'legal' and 'practical' responsibilities of management. Legal responsibility depends upon the policy of the statute in question: it may be imposed upon employers or managers, or both. When an employer is a corporate body, practical responsibility is always carried by managers, as they are the human agents through whom the corporation must comply with its legal duties.

The main criminal responsibility under the earlier statutory provisions was for the provision of resources necessary for the achievement of a safe working environment. Before the Act, the term 'working environment' was taken to mean buildings, machines and equipment. The old legislation contained mechanisms for 'strict' enforcement of its relatively limited requirements and imposed liability for failure to reach the necessary technical standards of safety (frequently expressed impersonally) upon the enterprise itself.

On the other hand, under the Act, responsibility for the organization and maintenance of a complete safe system of working imposes far heavier demands upon the business. It requires not only expenditure of resources upon plant and machinery, but also upon work people and their training, as well as on the organization of their tasks: a much more difficult and time consuming responsibility. In addition, the emphasis in the Act towards placing responsibility on persons as well as organizations has more than a technical significance for managers. It must serve to bring home sharply to all of them the onerous nature of the duties being imposed upon management, as well as upon the organization in which it functions. It is clear that these management responsibilities under the Act are of a much heavier nature than those under the earlier statutory provisions.

Employers' duties and line managers' duties contrasted

When the Act imposes obligations upon a corporation it is in reality imposing duties upon two separate categories of managers, as follows:

(a) employer or 'board room level manager' duties: these are the duties of officers who act within the 'brain area' of a corporation, formulating corporate strategies and providing the necessary resources for achieving the objectives of the corporation, including its legal liabilities

(b) managers' duties: the duties of officials employed as line managers who apply the corporate resources and carry out the day to day activities in accordance with policies laid down by the board of the corporation in order to comply with the law.

Under such criminal legislation as the Act, the employer as an individual, or the corporate employer as an entity, will be criminally liable for a failure to comply with standards laid directly upon it by legislation. A corporation may be regarded as either 'personally' or 'vicariously' liable for this failure. This is a somewhat artificial concept in most instances, but analysis may be necessary where a statute draws a distinction between corporate and personal duties in determining criminal responsibilities. Speaking very generally, a corporation may be deemed to be 'personally' responsible if a strict duty directly laid upon it (for example as occupier or employer) is not in fact carried out: while a corporation may be deemed to be 'vicariously' responsible if it expressly delegates to another person (such as a manager, official or other employee) the discharge of the same duty, and that person neglects it. In reality the basic liability of a corporation is not changed by the delegation of a strict duty to a subordinate within the organization. However, the same statute may go on to make some separate provision for the delegation situation by, for example, imposing a separate obligation upon the delegate himself, or by providing a mechanism for exempting the corporation from liability when an otherwise satisfactory system for compliance has been disrupted by the negligence of a delegate, or by the casual act of an employee. In this way, the 'strictness' of the duty imposed on the corporation is qualified by a defence.

Even where a strict duty is created, in practice delegation to managers and officials is essential within all organizations so that the law is in reality holding the corporate body or the employer, responsible for the results of teamwork by his team. There may, or may not, be an identifiable culprit within the team, perhaps only the results of cumulative slackness. In a case of strict liability, this does not matter: what matters is that the law will hold the corporation to account for failure to secure a result or measure up to a standard. Additionally the boardroom level managers may in appropriate cases incur a personal criminal liability for their actions or inaction. This important subject is dealt with separately on pp 125–26.

Management responsibilities under the earlier statutory provisions

Under the Factories Act 1961 the corporation as occupier authorized the necessary expenditure to safeguard machines and appoint staff to carry this policy out: in the event of failure, only the corporate entity was in fact prosecuted. Again, under the same Act, the occupier was subject to an obligation to keep floors clear of slippery substances so far as was reasonably practicable. A corporation normally discharged this obligation by setting up a reasonable system for cleaning the floors and then delegating its performance, ie the actual cleaning, to an employee. If that employee neglected this duty, the occupiers were liable under the Act, subject to the right of naming the negligent employee as an 'actual offender', within the terms of the special third party defence in s161 of the Act[1]. This defence enabled the enterprise itself to identify an actual offender: generally but not exclusively from among its workforce, and to escape liability. It is significant how rarely this defence has been evoked over the years in safety legislation, presumably because of the need for the defendant enterprise to demonstrate in court its own diligence as well as to demonstrate the wrongdoing of the named party[2].

The Mines and Quarries Act 1954, by contrast, imposed specific responsibilities upon designated members of management, in addition to the overriding responsibilities placed upon the enterprise itself. These designated managers themselves bore a responsibility for the discharge of their specific obligations by any subordinates to whom they had delegated them. In that Act the duties of owners were set out in s1 and those of managers in s2. An example of the distribution of responsibility, and therefore of liability, can be seen in the case of ventilation. Here the owner's duty was to 'make such financial and other provision' for ventilation equipment as may be necessary. The duty of the manager was to 'take such steps as are necessary to secure' that adequate ventilation was constantly produced. A duty was imposed on junior officials (deputies) to ensure that ventilation was provided in particular parts of a mine, and so on. Thus a wide general duty upon the owner was underpinned by a narrow specific duty upon a descending order of persons within the management hierarchies.

The distribution of responsibility under the Act

Under the Act, the basic responsibility for compliance with the general duties is imposed upon 'employers', and provision is made for obligations under the regulations to be imposed upon 'persons' in the widest sense. This power could lead to a development, which for most managers would be novel; namely the introduction of the concept of personal management responsibility, which developed in the mining industry, and spread into a number of other dangerous trades. Managers might encounter particular responsibilities under regulations imposed upon designated individuals in addition to the overriding responsibilities of the enterprise itself; and for those designated managers themselves to bear a responsibility for the discharge of their obligations by any subordinates to whom they delegated them[3].

The basic principle to remember where the Act and Regulations are concerned is that whenever a specific duty is laid on a designated 'person', be it corporate, employer or manager, that person remains criminally responsible to discharge the duty, even though it may be that the performance of that duty must be delegated to another manager, or perhaps a shop floor employee. Unless the manager can benefit from the defences in ss36 or 40[4], the designated person will be responsible for the failure to comply with the requirement concerned. It may be that in due course regulations will be made distributing different aspects of a duty among officials in a hierarchy, as in the ventilation in coalmines example we have considered, but delegation as such will be no answer to the charge of non-compliance with the particular requirements.

It should also be remembered that the Act, in contrast to the earlier statutory provisions, does not contain a third party defence. It is difficult to assess the importance of the omission at this stage, as so little use was apparently made of the defence under that legislation. It is true that s36 still permits an inspector to proceed directly against the actual offender, rather than the person upon whom the statutory duty is imposed; but this section grants a discretion to the inspector and gives the person primarily responsible no legal rights.

Management responsibilities and civil liability

A useful analogy with the new pattern of responsibilities of management can be found in the pattern of common law liability for injury to employees, with which employers and insurance companies will be familiar.

For the purposes of common law liability, it can be taken that managers, owners and workers are each under a duty to take reasonable care for the safety of their 'neighbours', but that effectively only the employer (the corporate enterprise) has the resources to meet a common law claim. Nevertheless, in strict law, every person within the organization owes a duty at common law in relation to safety. The basis of this duty is to maintain acceptable standards of safety at work, through the taking of positive steps so far as lies within an individual's power. The policy must be laid down at boardroom level and must be adequate to ensure that the physical environment of the workplace is of a reasonably safe standard and is maintained at that level; and that the workforce is recruited, trained, supervised and disciplined to a pitch of safety that is adequate to discharge the corporate obligation to function within and maintain the safety of the physical environment so as not to endanger other employees, visitors and the public.

The discharge of these obligations within a corporate organization is a matter of simple common sense as much as law. Let us suppose that a corporation acquires plant for the manufacture of chemicals by a hazardous process. The next step is to provide a staff 'competent' to operate this plant as safely as possible. For this purpose, it matters not in the eyes of the law whether or not the corporate members as such have any knowledge of the particular hazards involved, providing that they engage staff with this necessary specialized knowledge. The employer will be personally responsible if staff who lack this knowledge are appointed and a disaster ensues, on the grounds that the staff appointed are not 'competent', whether or not it can be established that the particular manager or official was 'negligent' in any personal sense[5].

Assuming that a manager who is fully competent in the legal sense is appointed, but neglects his duties, the employer remains liable, although on this occasion the responsibility takes the form of vicarious liability for the manager's personal negligence in the course of his employment. Assume further that the manager is competent, and is not negligent, but that he has delegated certain functions to his junior, and, in the event, the junior's negligent conduct is the cause of the accident. Virtually the same considerations apply; under the common law, the employer is responsible for the negligence of the employee concerned.

The 'risks versus costs calculus'

Duties under the compensation law are to do what is reasonable in all circumstances. In determining what was reasonable, the courts had built up (prior to the imposition of criminal liability under the

Act) a calculus, as it were, in which the risk to the employees that the law sought to minimize by the requirement in question was weighed against the cost of the precautions to be taken, whether in money, time, loss of business and the like; the question of liability has been determined from this calculation. For example, if a risk is grave and the required precautions not particularly costly in this sense, it would be reasonably practicable to comply with them; equally there are situations where the reverse will be found to be true. This common law calculus can give most helpful guidance on the likely requirements under the Act to comply with the general duties.

Sometimes, however, in statutes the defence is that of practicability, without qualification; here the test will not be so much the 'cost' of the precautions but whether they can be achieved in an engineering or scientific sense – clearly a different calculation altogether, imposing far heavier obligations.

'Qualified' duties under the Act

It is suggested that management must approach its criminal responsibilities under the general duties and under the health and safety regulations in this light. Obviously, what is practicable or reasonably practicable for a named manager to achieve will depend very much on his status within an organization, his powers and his control over its resources. For instance, if a new process requires safeguards costing £10,000, it is reasonably practicable for an employer to provide them. But let us suppose that a like obligation is placed upon the manager in future regulations: would a manager be responsible if the £10,000 were not provided? Financial problems of this type have required elaborate defence provisions in the Mines and Quarries Act 1954. The scheme of the Factories Act placed basic responsibility for all breaches, other than disciplinary offences, upon the enterprise as occupiers, preventing this problem from arising. If duties under the Act relating to the expenditure of funds, or allocation of resources, are to be imposed upon members of management, provision will have to be made in regulations under s15 for the protection of these individual managers.

Problems of a different nature may arise in the serving of inspectors' notices. The improvement notice will require the person made responsible under the Act, or regulation in question, to take steps to comply with the law. In many instances the employer will be the person criminally responsible, but the necessary action to secure compliance will in practice be taken by managers. It does not appear that the manager will incur primary criminal responsibility if an improvement notice is not obeyed, but possibly he may be guilty of an offence under s7 or s36 if the failure to obey is due to his neglect to do all within his power to comply. Prohibition notices, on the other hand, lay obligations on the person in control of the activity in question. The essence of the offence in this instance is failing to cease operations in compliance with the notice. It is difficult to see how a manager could incur criminal liability for breach of a prohibition notice if he lacked powers of control sufficient to order the cessation of the activity, as required by the notice.

Management responsibilities and safe systems of working

A vexed issue may well arise in requiring discharge of an employer's or manager's duty so far as is reasonably practicable, if the method chosen for the discharge of the duty is to secure compliance from the employees in some way. For example, the Act may require certain workers to be protected against inhaling dangerous dusts, so far as is reasonably practicable. An employer or manager may choose to comply with this obligation by the issue of dust masks, instructing and exhorting the workers concerned to wear them. Management must weigh the problems arising from enforcing this requirement (discussed in chapter 3) upon the employees against the risks of being prosecuted by the Inspectorate, or served with an improvement or prohibition notice if the workers' health is not in fact safeguarded in this way. A likely source of failure would be inability to enforce the wearing of these masks by employees generally. It is doubtful whether an employer faced with a prosecution or notice under s2 could plead successfully that is was not reasonably practicable to do more to enforce the wearing of masks by employees concerned[6].

Responsibility and delegation

Where a manager delegates a task, he may still be obliged to establish that he has discharged his duty so far as is reasonably practicable under s40. The burden of proving that an action was reasonably practicable for the purpose of this section, when related to s36, is very heavy. In attempting to discharge an obligation under s40, a defendant would usually need to satisfy a court that it was not reasonably practicable, not only for himself but for his employees or other persons to whom he had delegated the taking of precautions, to have done more. It follows from this that it would not be a defence within s40, whatever may be the position under s36, for an employer to prove that he himself had done all that was in his power to comply, but that he had been let down by an employee, or subordinate, or senior member of the management chain.

Criminal liability of managers under the Act

(a) Managers at boardroom level

Where an obligation is laid upon a corporate employer or upon a 'person', which is in fact a corporation, it is clear that the offence is committed by the corporation, and not primarily by the members of the board whose acts or omissions have indirectly or directly caused the offence to be committed. However, s37 of the Act contains a fairly standard director's liability clause imposing criminal liability in addition upon any director, manager, secretary or other officer of the firm, whose consent, connivance or neglect has caused or contributed to the offence in question, so that the officer, as well as the firm shall be guilty of an offence and shall be liable to be proceeded against and punished. The Factories Act 1961 contained a similar provision, but it does not appear that any prosecutions had been brought against corporate officers under it. It is not easy to identify any changes in the responsibilities of management at boardroom level brought about by the Act that will render prosecutions under s37 more likely than under the provisions of the Factories Act; nevertheless s37 has been used on a number of occasions. A controversial conviction under s37 was that of the Director of Roads for Strathclyde in respect of a fatal accident which was deemed to have occurred as a result of the Director's failure to implement his employer's safety policy, in contravention of the duty which the safety policy itself imposed upon him[7]. In another case a director of the Singer sewing machine company was convicted under this section even though no charges were laid against the company itself[8].

There is a possibility that board room level managers might be proceeded against, in a similar situation under s36. Again the likelihood of this course of action would depend upon whether the fault that it was sought to bring against the named person was deemed 'personal' to the corporation. It might well be argued that, inasmuch as the manager represented the company by his actions, his personal fault was in fact the fault of the company[9]. But it seems reasonable to assume that, providing an offence had been committed through the agency of the particular manager, the inspector would be able to succeed against him under either s36 or s37.

(b) Line managers

A line manager may commit a criminal offence under the Act in a number of ways. First, where an obligation is laid upon a corporate employer but its discharge is necessarily delegated to the manager, he may be guilty of an offence under s7 or 36 (*see* chapter 9). The point may be illustrated by the instance of an unfenced machine in a factory. The basic obligation to fence under either s14 of the Factories Act or s2 of the Act will fall upon the occupier or the employer respectively. This may well mean in practice that the duty is imposed upon a corporate body. Responsibility for safeguarding the machinery within the organization under the corporate safety policy will be assigned to a particular manager. If the latter neglects this obligation, he may be prosecuted under s7(b) as an employee who has failed to 'cooperate' with his employer to enable the latter to discharge his duty to safeguard the machine[10]. Alternatively he may be prosecuted under s36 as 'another person' without necessarily relieving the corporation of its primary responsibilities. If on the other hand the manager were to delegate this duty, in a responsible manner, to an entirely suitable and competent person such as a qualified safety officer, and if the safety officer were to neglect his duties, both the corporation and the safety officer (as 'another person') may well be guilty of an offence, but not the manager himself.

Secondly, under regulations an obligation may be imposed upon a manager 'personally'; here, criminal liability depends upon the wording of the regulation. If the obligation is 'strict' a manager, like an employer, could be held to account if it were not complied with, either on a 'personal' or vicarious basis. If a manager delegates the performance of such a duty to others he would retain primary responsibility, so that if the delegate in his turn failed to comply with the requirement, both the manager and delegate would be criminally liable and might be prosecuted. A manager's criminal liability in this situation is quite different from his liability as a delegate, as discussed in the previous paragraph. Where a manager's failure to comply with the requirements of the Act is due to a technical failure, such as a latent defect in a machine, as opposed to a human failure, s36 would have no relevance. In this instance the criminal liability of the manager would depend on whether that liability was qualified by reference to practicability, or reasonable practicability; or was 'strict' (similar to fencing of machinery requirements under the Factories Act). Only if the requirement was qualified in some way could the accused plead the failure of the machine or equipment; but of course, even where a strict requirement is breached, the failure of a machine could be pleaded in mitigation of penalty.

Strict offences, vicarious offences and the criminal courts

As a general rule, the criminal courts will not convict an accused person in the absence of clear proof of forbidden acts, coupled with a criminal intent. An exception has been made to this rule in the case of certain comparatively minor offences created by statutes, such as the earlier health and safety legislation. In this instance penalties have been imposed on employers (under the Factories Act) and owners and managers (under the Mines and Quarries Act) for what were essentially either latent defects in machines or neglect by subordinates of the duties delegated to them in the course of their employment.

This exception to the general rule has been increasingly criticized in recent years. In fact the House of Lords has stated[11] that courts should be reluctant to convict in the absence of guilty intent, even when enforcing public welfare legislation such as that on health and safety. This is particularly so in two instances (a) where the person accused lacked the 'power to prevent the illegality', and (b) where the Act in question created 'truly criminal' offences. Both of these criteria are relevant to the criminal liability of managers under the Act[12].

Criminal liability and the 'power to prevent'

An employer, and a line manager acting on his behalf, are deemed to have control over the employees in the course of their employment: thus it is not easy to argue, when an offence has been committed under the Act through the neglect of a subordinate, that it was beyond the power of the accused employer or manager to prevent the breach. It is quite irrelevant to argue that the subordinate may have committed a separate offence himself under s7, or may be proceeded against by an inspector under s36: if a duty is laid upon a manager or employer, which he chooses to discharge by delegating to what turned out to be an unreliable subordinate, it cannot be maintained that the delegator is excused on the grounds of lacking power to prevent the contravention. Nor can it be argued that an accused person lacks the power to prevent a contravention, when what is really meant is that prevention would have been difficult or expensive, if it in fact could be done (in the sense of being within the powers of that person) at however great an inconvenience to himself. In a decision of the House of Lords a firm accused of causing polluted matter to enter a river claimed that it had no power to prevent the pollution when its effluent recovery system overflowed into the river. It was the view of their Lordships that the company certainly possessed the power to prevent the offence, because it could have arranged for the automatic stopping of its operation whenever the effluent system came near to overflowing[13].

The power to prevent could well be an element in determining liability on a charge of failure to comply with an inspector's order. This point has already been touched on in this chapter: an example of circumstances in which this might be relevant would be where an accused person attempted to avoid liability for contravention of a prohibition notice by pleading that he did not have power within an organization to order the shut-down of premises or of a process in compliance with the notice.

'True crimes'

It is doubtful whether the summary offences created under the earlier statutory provisions would have been regarded as 'true crimes' by the House of Lords. But there can be little doubt that the Act's provisions for trial on indictment carrying the possibility of unlimited fines or imprisonment would fall into this category. It would be entirely in accordance with the Robens philosophy if under the Act criminal prosecution were resorted to by an inspector (as an alternative to the issue of a notice) only in cases of 'flagrant, wilful or reckless' disregard for safety and health[14]. Enforcement statistics since the introduction of the Act indicate that relatively few cases have been brought upon indictment, but a considerable number of improvement and prohibition notices have been served, while the general level of prosecutions has remained much the same as before the Act. Hence the power to take proceedings upon indictment under the Act is in practice exercised only in cases of personal and intentional actions or omissions contrary to the Act, or where contraventions have led to heavy loss of life.

It is impossible to imagine that a sentence of imprisonment would be imposed on a defendant who had not been convicted of a deliberate intention to ignore the obligations laid upon him by the Act. In fact very few offences that a manager might commit under the Act carry this penalty. Moreover there are apparently no precedents in English law for a sentence of imprisonment for an offence of vicarious liability, and few indeed for offences of negligence, as the history of convictions for the most serious driving offences under the Road Traffic Acts makes clear.

A manager considering his heavy responsibilities under the Act can at least rest content that he does not face the prospect of imprisonment except for criminal offences involving a wicked indifference to, or defiance of, his obligations for the health and safety of others. But it should be noted carefully that defiance of a prohibition notice is a criminal offence where the penalty of imprisonment may be imposed, and it is suggested that this is a situation in which a court might well consider the heavy penalty appropriate.

Management responsibilities and judicial attitudes

The special expertise of juries is in assessing people rather than weighing technical factors. In the past, juries have been alleged to be sympathetic to 'white collar' crimes generally, and particularly understanding where inadvertance and negligence have led to prosecution for offences under statutes such as the Road Traffic Acts. Emerging case law under this Act indicates that management who are in violation of occupational health and safety laws are enjoying the benefit of this traditional approach. Moreover, the determination of the penalty to be exacted if a verdict of guilty is returned is a matter for the judge, and it may be that there is a discernible judicial reluctance to impose high penalties in an area of law which has previously been associated with strict liability and moderate penalties.

12 Protection of the public

The problems

There are many work activities that can give rise to hazards which would affect the safety of the public. Some of these occur frequently but only create small scale risks, for example, construction work adjoining a highway. Very large scale risks, where accidents could reach disaster proportions, are created very much less frequently. Situations of potential risk can be created either by entirely new developments or simply by increasing the size and scale of existing operations. Problems can be created by accumulating potentially hazardous situations through the activities of a number of separate organizations operating in close proximity.

The reality of these hazards was emphasized by a number of catastrophes in the 1960s, such as the Brent Cross crane accident[1], the Aberfan disaster[2] and the Hixon level crossing accident[3]; health hazards were underlined by a number of reports of neighbourhood poisoning in the vicinity of industrial plants[4]. The Report of a public enquiry into the crane accident at Brent Cross recommended that consideration should be given to amending the Factories Act so as to bring the public directly within its scope. Similarly the tribunal appointed to investigate the cause of the Aberfan disaster recommended that the Mines Inspectorate should be required to consider the safety, health and welfare of persons in the vicinity of a mine.

The problem of the traditional approach to occupational health and safety was that it did little[5] to protect the public against risks to which they might be exposed as a result of the activities of persons at work. The earlier statutory protection against occupational hazards had the effect of drawing a somewhat artificial distinction between care for the safety of workpeople and care for the safety of the general public. The legislation was not framed so that it could provide direct protection for the public; nor had inspectorates any jurisdiction regarding public safety when they went about the task of administering and enforcing legislative provisions. No obligation existed to record, report, or investigate injuries to the public caused by work activities.

The Robens proposals

The Robens terms of reference included an instruction to consider whether any further steps were required to safeguard the public from hazards other than general environmental pollution arising in connection with activities in industrial and commercial premises. A major concern expressed in the Report was the relationship between legislative provisions enacted to achieve occupational health and safety, and the safety of the public generally[6].

The Report expressed concern that there was no effective system for preventing the creation of large-scale risks to the public, simply because there was no effective mechanism for controlling the siting of major industrial installations at which potentially hazardous industrial activity might take place. It was noted that a marked feature of industrial development in recent years had been the increased storage and use of some intrinsically dangerous substances with highly explosive, flammable or toxic properties.

Local authorities charged with granting planning permission were concerned with controlling the class of land user for the preservation of public amenities. They had no duty to prevent the storage of potentially dangerous chemicals, to control processes, or indeed the scale of production. They had no duty to monitor the operation of plant once its presence had been sanctioned for planning purposes, provided the user of the property remained within the same broad class. Had they wished to take these factors into account when considering planning applications, they might well have lacked the technical expertise to identify the risks. These problems were becoming increasingly pressing as the scale of industrial activity increased and as the storage of chemicals of large quantities became more common. Nor did the Factories Act provide a safety net, for generally the introduction of new processes (as distinct from a change in the occupancy of the premises) did not have to be notified to the Factory Inspectorate.

The Committee recommended that statutory safety provisions,

while designed in the first instance for the protection of employees, should be so framed and administered as to enable full account to be taken of the related public safety considerations. Regulation-making powers should be wide enough to permit regulations protecting the public against 'internal' and 'external' risks (to persons on or near the premises in question). A special unit was needed in the proposed safety authority to concern itself with major hazards and to give advice to local authorities and industrialists.

The recommendations concerning Major Hazards were accepted prior to the Act by the Government. The Major Hazards Coordinating unit was set up and, in due course, the Major Hazards Branch was established within the Health and Safety Executive. A fact-finding and advisory Committee of Experts on Major Hazards was also established, whose membership was drawn from universities, industry, research institutes and public services. This Committee had the following terms of reference:

> To identify types of installations (excluding nuclear installations) which have the potential to present major hazards to employees or the public or the environment, and to advise on measures for control, appropriate to the nature and degree of hazard, over the establishment, siting, layout, design, operation, maintenance and development of such installations, as well as overall development, both industrial and non-industrial, in the vicinity of such installations.

The early work of the Committee was to identify major potential hazards and propose that plants which are so identified be classified as 'notifiable installations'[7].

The Major Hazards Branch subsequently underwent reorganization; the Hazardous Installations Group with two branches was set up. The first branch is the Hazardous Installations Policy Branch (which has similar responsibilities to those of the Major Hazards Branch) including servicing the Advisory Committee on Major Hazards, and devising regulations governing 'notifiable installations'.

The second branch is the Major Hazards Assessment Unit. This is responsible for developing methods of assessing risks and identifying areas where codes of practice would be helpful. It is charged with advising the inspectorates on how to deal with difficult safety problems arising out of hazard surveys produced by occupiers of particular installations.

The Flixborough disaster

In June 1974 (while the Health and Safety at Work Bill was before Parliament) the Flixborough plant was virtually demolished by an explosion which killed 28 and injured 35 at the workplace. Outside the works, 53 people were recorded as casualties, and hundreds more suffered relatively minor injuries which were not recorded. Property damage extended over a wide area and a preliminary survey showed that 1,821 houses and 167 shops and factories had suffered to a greater or lesser degree[8].

This major disaster demonstrated the validity of the Robens prediction that large scale industry could create major hazards that endangered not only the workforce but also the surrounding community.

Legislative provisions

Section1(1)(b) of the Act states that the provisions of the Act shall have effect with a view to 'protecting persons other than persons at work against risks to health and safety arising out of, or in connection with, the activities of persons at work'. The effect of this provision is to permit regulations to be made for the specific purpose of protecting the public. It also enables the inspectorates to exercise their powers with a view to the protection of the public.

Section 1(1)(c) makes provision for controlling the keeping and use of explosives, and s1(1)(d) makes provision for controlling the emission into the atmosphere of noxious and offensive substances. These provisions may be of particular relevance to public safety. As far as s1(1)(d) is concerned it is important to bear in mind that the former Alkali Inspectorate has been brought within the Executive.

Sections 3 to 5 impose duties which are wholly or incidentally concerned with the protection of the public.

Section 3: duties of employers

Under s3(1), employers are placed under a duty to conduct their undertakings in such a way as to ensure, so far as is reasonably practicable, that persons not in their employment who may be affected thereby are not exposed to risks to their health and safety. There is nothing in the wording of this section to restrict its protection to persons at work; both visitors to, and neighbours of, a hazardous installation come within its protection. It is clear that this duty is wide as well as novel. The following examples indicate the scope of an employer's current obligations to the public:

(a) an employer who stores hazardous or inflammable substances on his premises must take into consideration the risk to the surrounding public in the event of a fire[9]

(b) the obligation also extends to the transportation and handling of toxic and hazardous substances in the course of conducting an undertaking. Regulations have recently been made which seek to protect the public, transport users and other road users against the hazards of improperly labelled or packaged substances[10]

(c) the public protected by these provisions may be either on or off the employer's premises. Thus the occupier of a construction site will discharge his obligations by excluding the public from his premises and preventing falls of material beyond the perimeter; at the other extreme a hospital, school, or theatre must take steps to safeguard the public that are invited onto its premises for any purpose

(d) finally, it should be noted that the protection extends to health. It can be argued that dangerous dust or excessive noise emanating from a workplace would injuriously affect the health of persons living in the vicinity and thus contravene s3(1).

Section 3(2) places a similar obligation upon self-employed persons. Section 3(3) enables regulations to be made requiring employers and self-employed persons to provide information about such aspects of the way in which they conduct their undertakings as might affect the health and safety of others. Again this provision is wide enough to require disclosure of information to the general public where necessary for their protection. It appears that this provision does not limit the duty which employers owe to the public under s3(1) to provide information where it is necessary for safety, but relates only to the issue of specified information required by regulations which may be made under s3(3). It is interesting to note that a recent EEC Directive requires employers to inform persons, liable to be affected by a major accident originating in a notified industrial activity, of the correct behaviour they should adopt for their own safety in the event of a major accident[11].

Section 4: duties of controllers of premises

Section 4 imposes a duty upon controllers of premises for the protection of persons visiting their premises. This duty, to take such steps as are reasonably practicable to ensure the safety of visitors could relate to visiting members of the public as well as visiting workmen. The criteria determining liability are that the visitor must be either a worker or else a person using equipment or substances provided on the premises for his use.

Public parks, playing fields and playgrounds are all covered by this section. Where persons are employed therein, s3 is also applicable, but s4 must be observed even where there are no persons regularly employed. The duty extends to persons who have control of premises to any extent. It is possible to envisage situations where more than one organization may owe a duty under this section (local authorities and playing field organizations may both have some measure of control, for the purposes of the section, over playing fields or playgrounds).

Section 5: emissions into the atmosphere

Section 5 imposes obligations designed to prevent emissions of noxious or offensive substances into the atmosphere and to render harmless and inoffensive those substances that are emitted. Emissions endangering the safety or health of the public also fall under s3(1): the provisions of s5 are somewhat wider in that they extend to emissions which are offensive and objectionable on environmental grounds.

Regulations protecting the public

The general purposes of the Act are sufficiently wide to permit the making of regulations on any subjectmatter related to the protection of the public from risks created by the activities of persons at work. It is possible that this regulation-making power may, upon occasion, be employed to create specific licensing provisions aimed at the controlling of the use of premises in the interests of public safety.

The enforcement agencies

The general purposes of the Act give the Executive power to carry out its enforcement functions with a regard to the problems of public safety. The Major Hazards Branch of the Health and Safety Executive has particular responsibility for the vetting of legislation and regulations from the aspect of public safety. It is also particularly concerned with specific licensing proposals. Within the Major Hazards Branch is the Risk Appraisal Group, which has the function of liaising with local authorities who care to approach it on a voluntary basis. If a local authority receives a planning application which appears to have implications for public safety, it may approach the Group which will offer advice and, if necessary, attend and give expert evidence at a local planning enquiry. This is a continuation and extension of a scheme devised by the Department of Environment and the Factory Inspectorate in 1972, under which local planning authorities were advised to consult the Factory Inspectorate before granting planning permission for development which might involve the use and storage of dangerous substances[12].

Growing importance of public safety

Managers evaluating their obligations to employees would be wise not to disregard the positive duty to protect the public. This duty is rapidly assuming increasing importance as public concern about the problems of the environment builds up and the inspectorate becomes familiar with the new range of its duties. Increased public awareness of the danger to them, created by the work activities of others[13], and the nature of the establishments of many of the new entrants (colleges, hospitals and hotels) has resulted in a high proportion of resources of the Health and Safety Executive being devoted to matters of public safety[14]. This emphasis upon public safety is certainly in keeping with the contemporary philosophy of promoting occupational safety in the widest sense.

CONCLUSIONS

In the 1974 Act Parliament attacked the problems of health and safety from a broad, common sense, viewpoint. This approach demands a positive response from management, not in terms of a checklist mentality but of an attitude of commitment to the ideal of making their enterprises safer and healthier for all concerned.

This response is demanded from every, and not a minority of, employers, not just from those managing mines and factories, but from all who control organized economic and social activities involving 'work'. It must not be forgotten that the Act applies as much to the family business as it does to the nationalized industry or government department; to shops, offices, laboratories, schools and hospitals, as strictly as it applies to industrial plants.

Now that the final stage of the scheme of the Act has been reached, by bringing into force the provisions for the appointment of safety representatives from employees, it is unlikely that the problems of achieving and maintaining safety at the workplace will ever be far from the personnel manager's mind.

While it cannot be denied that safety is extremely costly, it nevertheless seems likely, that as the national philosophy moves further from regarding financial gain as the sole criterion of successful industrial activity, the general health and safety of the community will become an increasingly important objective in the evaluation of industrial activity.

References and notes

Chapter 1

1 'Visitors' and 'neighbours' are phrases well known to lawyers handling personal injury compensation claims; *see* chapter 3

2 Regulations under the Mineral Workings (Offshore Installations) Act 1971 have similarly identified the personal responsibilities of management

3 In the event the Act did not expressly exclude this class of worker

4 *Report of the Committee on Safety and Health at Work*, 1972

Chapter 2

1 The Mineral Workings (Offshore Installations) Act 1971, and other legislation relating to activites on the British part of the continental shelf were, for technical reasons, kept outside the scope of the Act

2 Defined in s52. At the time of going to press only domestic employment (s51) is expressly excluded

3 The armed forces are, and the police force may be, outside the scope of the Act; however the Act has been extended by Regulations to apply to certain offshore installations outside British territorial waters

4 See eg The Notification of Accidents and Dangerous Occurrences Regulations 1980 which introduced a common system of accident reporting at work from 1 January 1981

5 'Asbestos Work on Thermal and Acoustic Insulation and Sprayed Coatings.' First Report of the Advisory Committee on Asbestos (1978)

6 But see Woodworking Machines Regulations 1974

7 Section 55 of this Act requires the Equal Opportunities Commission 'to keep under review the relevant statutory provisions in so far as they require men and women to be treated differently'. See the EOC report on this subject *Health and Safety Legislation: Should we distinguish between men and women?* (1979)

8 Croner's *Health and Safety at Work* and *The Industrial Relations Review and Report* are of especial value to personnel managers: in particular, in the latter context, its *Health and Safety Information Bulletin* (HSIB)

9 There are 32 existing statutes, with the inclusion of the Baking Industry (Hours of Work) Act 1954, by virtue of the Sex Discrimination Act 1975

10 *Health and Safety, Report on Manufacturing and Service Industries 1979*

Chapter 3

1 The Factories Act itself did not, of course, protect employees in such situations. See eg *Close* v *Steel Company of Wales Ltd* (1962). A list of all cases cited, with full references, is given in Appendix III

2 See *R* v *Swan Hunter Shipbuilders Ltd* (1981) on s2 and s3

3 Technically such an action may lie for breach of contract by the employee. See *Lister* v *Romford Ice and Cold Storage Co Ltd* (1957) and *Janata Bank* v *Ahmed* (1981)

4 See in particular *Alidair Ltd* v *Taylor* (1974)

5 Eg *Pagano* v *HGS* (1976) (unsafe delivery van); *Keys* v *Shoefayre* (1978) (unsafe premises); *BAC* v *Austin* (1978) (unsuitable goggles); *Oxley* v *Firth* (1980) (temperature in tailoring)

6 See judicial comment to this effect in *R* v *Swan Hunter* (1981)

7 Laid down by Parker J in *Adsett* v *K and L Steelfounders and Engineers Ltd* (1953)

8 See *Marshall* v *Gotham Co Ltd* (1954)

9 This burden may be discharged by proving, on balance, the

probability of the matters in question: see *Public Prosecutor* v *Yuvaraj* (1970)

10 Eg as in *Nimmo* v *Alexander Cowan & Sons Ltd* (1967)

11 For criminal proceedings under the Act see chapter 7

12 It is debatable whether an appeal against a notice is 'proceedings for an offence' under s40

13 An employee was prosecuted under s7 following a fatal accident occurring during working hours, while he was driving his own vehicle on his employer's premises (*see* HSIB March 1978)

14 Dangerous Substances (Conveyance by Road in Road Tankers and Tank Containers) Regulations 1981

15 *R* v *Swan Hunter* (1981)

16 See the problem posed in relation to civil liability for scrotal cancer, *Stokes* v *GKN Ltd* (1968)

17 *Stokes* v *GKN Ltd* (1968)

18 At present regulations exempt employers with less than five employees in an 'undertaking' Employees Health and Safety Policy Statements (Exceptions) Regulations 1975. See *Osborne* v *Bill Taylor of Huyton* (1982); held that (i) 31 branches of a shop might, for these purposes, be treated as one undertaking, but that (ii) the exemption might extend to an undertaking where more than five employees were on the pay roll if fewer than five were on the premises at any one time

19 See *Fire on HMS Glasgow 23 September 1976*

20 But see *J Armour* v *Skeen* (1977)

21 See generally the HSE publication *Effective Policies for Health and Safety*

22 At the time of writing (March 1982) no such regulations have been issued

23 A consultative document on the general subject of protection of employees against noise at work was issued by the HSE in 1981

24 The comparable US legislation, the Occupational Safety and Health Act 1970, makes no attempt to impose criminal liability upon employees

25 Eg a 62 year old worker was put in a lift designed to carry only goods. The lift first jammed then plunged to the bottom of the shaft; its occupant suffered a broken back. The two men who did this were prosecuted under s7 (HSIB March 1978)

26 Following the Houghton Main mining disaster senior management in the mine were prosecuted under s7. (See *Report on Explosion at Houghton Main Colliery, Yorkshire*) (1976)

Chapter 4

1 *McArdle* v *Andmac Roofing Co* (1967)

2 *R* v *Swan Hunter* (1981)

3 A prohibition notice was issued in 1978 to prevent the incineration of the insecticide and fungicide kepone in order to protect both workers and public living nearby

4 *R* v *Swan Hunter* (1981)

5 *Wheat* v *E. Lacon & Co Ltd* (1966)

6 *General Cleaning Contractors Ltd* v *Christmas* (1952) (window cleaner); *Smith* v *Austin Lifts Ltd* (1959) (maintenance engineer)

7 It has been suggested that where premises are under the exclusive control of a sub-contractor liability for defective access rests primarily on the party controlling access under s4, rather than on the conductor of the undertaking under s3. Thus some delimitation between the two sections may be emerging (*Aitchinson* v *Howard Doris Ltd* (1979))

8 Robens Report, para 330

9 See *Vacwell Engineering Co Ltd* v *BDH Chemicals Ltd* (1970) – duty to customers; *Cassidy* v *Dunlop Rubber Co Ltd* (1972) – duty to customers' employees.

10 See *Articles and substances for use at work – guidance for designers, manufacturers, importers, suppliers, erectors and installers.* (HSE Guidance Notes GS/8)

11 As interpreted in s53

12 Compare the duties owed by controllers of premises in respect of articles and substances under s4 (see pp 35–38)

13 The regulations are applied to consumer transactions by authority of the European Communities Act 1972, s2 and to employment situations by HASAW Act s6

14 Motor Vehicles (Construction and Use) Regulations, 1963

Chapter 5

1 It is intended to create a number of Advisory Committees with specialist knowledge to assist HSC and industry in these matters: approximately 10 of these committees, many of them in respect of specific hazards, have now been created (*see* chapter 2)

2 See *Advice to Employees, Guidance Notes* HSC Series 5

3 It is an offence falling within s33(3)

Chapter 6

1 See eg, *The Diving Operations at Work Regulations* 1981 which provide examples of the topics marked in the text

2 *Ibid*

3 *Ibid*

4 The Health and Safety Executive is at the time of writing exercising the power to licence nuclear sites under the Nuclear Installations Act, and to license key personnel under the Mines and Quarries Act

5 See regulation 5 of *The Control of Lead at Work Regulations 1980*, and paragraph 21 of *The Control of Lead at Work, Approved Code of Practice 1980*

6 See *Consultative Document on Notification of New Hazardous Substances 1981*

Chapter 7

1 This technical issue held up the hearing of many cases in 1976 and 1977. See *Campbell* v *Wallsend Slipway and Engineering Co Ltd* (1977) and *Robson* v *Brims & Co Ltd* (1977)

2 See ss7, 36 and 37

3 The Industrial Tribunals (Improvement and Prohibition Notices Appeals) Regulations 1974 No. 1925

4 See *Otterburn Mills Ltd* v *Bulman* (1975)

5 *A C Davis & Sons* v *The Environmental Health Dept of Leeds City Council* (1976)

6 *Belhaven Brewery* v *McClean* (1975)

7 *T C Harrison (Newcastle under Lyme) Ltd* v *Ramsey* (1976)

8 As in *Nico Manufacturing Co Ltd* v *Hendry* (1975)

9 Industrial Tribunals (Improvement and Prohibition Notices Appeals) Regulations 1974

10 *Chrysler UK Ltd* v *J D McCarthy* (1978)

11 This power was exercised by the Barnsley magistrates in a prosecution of the British Steel Corporation following an accident at Appleby Frodingham works in which seven employees had been killed by an explosion of molten metal

12 See *Tesco Stores Ltd* v *Edwards* (1977)

13 See Lord Salmon's criticism of the former factory inspectorate in *Dodd* v *Central Asbestos Co Ltd* (1972)

14 See *Dutton* v *Bognor Regis UDC* (1971); *Anns* v *Merton* (1978)

Chapter 8

1 *Secretary of State* v *ASLEF* (1972)

2 For first aid generally see The Health and Safety (First Aid) Regulations 1981

3 *Boyle* v *Kodak* (1969)

4 *White* v *Pressed Steel* (1980)

5 In the case of an employee, this would normally be s7, or possibly some duty imposed by new regulations or under the earlier statutory provisions

6 In some cases involving a great disregard for safety it has been held that the employee has by his action terminated his own contract (see *Gannon* v *Firth* (1976)) but the courts are nowadays disinclined to adopt this view and require the employer to dismiss the employee

7 *Turner* v *Goldsmith* (1891)

8 *Langston* v *AUEW* (1974)

9 *Spencer* v *Paragon Wallpapers Ltd* (1977)

10 *Kenna* v *Stewart Plastics Ltd* (1978) (epilectic causing distress to fellow workers)

11 Eg under *The Ionizing Radiations (Sealed Sources) Regulations* 1961

Chapter 9

1 *R* v *Swan Hunter*

2 Section 37. *See* chapter 10 for a discussion of management responsibilities under the Act

3 Quarries (Explosive) Order 1959; Considered in *ICI* v *Shatwell* (1964)

3. See chapter 10 on safety representatives

5 *Pagano* v *HGS* (1976); *Mayhew* v *Anderson (Stoke Newington) Ltd* (1977); *Kaye* v *Shoefayre Ltd* (1978); *BAC* v *Austin* (1978)

6 See *St Anne's Board Mill Co Ltd* v *Brien* (1972)

7 See Employment Protection (Consolidation) Act s64 and Schedule 13

8 A campaign by an individual would be unlikely to be protected trade union activity (See *Gardner* v *Peeks Retail Ltd* (1975)) even if the worker involved were a union member (*Chant* v *Aquaboats Ltd* (1978))

Chapter 10

1 Additionally the Coal Industry Nationalization Act 1946 has placed a statutory obligation upon the National Coal Board to consult its employees on matters including health and safety at work. See also similar requirements in the Electricity Act 1948 and the Iron and Steel Acts 1949 and 1967

2 It is sometimes suggested that safety representative work may be classified as 'trade union activities' within the meaning given to that expression under the Employment Protection (Consolidation) Act 1978 and therefore safety representatives are entitled to protection against discrimination under s23. See chapter 9. For this, and other reasons, some trade unions have considered that it may strengthen the position of a safety representative if he is an 'official' of the union within the definition of official used in a union's rule book. The lack of clear provision in these matters is in contrast with the equivalent US Occupational Safety and Health Act's clear treatment of safety representatives as performing specially 'protected' activity

3 The safety representatives' functions extend to 'welfare' so far as this is the subject of health and safety regulations or of any of the earlier statutory provisions

4 *Waugh* v *British Railway Board 1980*

5 HSIB No 45 September 1979

6 *White* v *Pressed Steel* (1980)

Chapter 11

1 As in *Braham* v *J Lyons Ltd* (1962)

2 Contrast the success of the retail trade in invoking a similar defence under s24(1) of the Trades Descriptions Acts 1968 and 1972; *Tesco Supermarkets Ltd* v *Nattras* (1971)

3 See, eg *The Diving Operations at Work Regulations 1981*

4 These defences are discussed in chapter 7

5 This point is illustrated by the findings of the *Report on the Flixborough disaster* (paras 19–25)

6 Following the disaster at the Appleby-Frodingham Steelworks in 1975 British Steel were convicted for failure to enforce their system for the use of protective clothing by blast furnace workers (Case unreported)

7 *J. Armour* v *J. Skeen* (1977)

8 HSIB 1978. NB: The director was convicted even though no charges were laid against the company itself

9 See *Tesco Supermarkets Ltd* v *Nattras* (1971) for a discussion of this highly technical issue

10 See eg the prosecution of members of management following the Houghton Main Colliery explosion in 1975

11 *Sweet* v *Parsley* (1969)

12 On the power to prevent: when the British Steel Corporation were prosecuted (see 6 *supra*) an uneasy compromise was reached. The corporation were convicted, though apparently lacking any guilty intention, but were only fined £700

13 *Alphacell* v *Woodward* (1972)

Chapter 12

1 See official Report, 1964

2 See official Report, 1967

3 See official Report, 1968

4 See eg Report on Lead Poisonings at the RTZ Smelters Avonmouth 1972

5 See however, Offices, Shops and Railway Premises Act 1963, s28: Fire escapes and protection of the public. Agriculture (Safety Health and Welfare Provisions) Act, 1956, s7: Agriculture (Avoidance of Accidents to Children) Regulations 1958: Mines and Quarries (Tips) Act 1969 (passed after Aberfan)

6 See Report, chapter 10

7 There have been two reports

8 See Report of the Court of Inquiry into the Flixborough Disaster 1975

9 See, The Fire and Explosions at Permoflex Ltd, Trubshaw Cross, Longport, Stoke on Trent, 11 February 1980

10 Hazardous Substances (Labelling of Road Tankers) Regulations 1978 Packaging and Labelling of Dangerous Substances Regulations 1978

11 EEC Council Directive on Major Accident Hazards from Certain Industrial Activities (December 1981)

12 Guidance as to dangerous substances was given to local authorities in a Department of Environment Circular (1/72): Development involving the use and storage in bulk of hazardous material

13 See Canvey Island Report

14 See, eg the following Pilot Studies by the HSE: *Working Conditions in Universities* 1978; *Working Conditions in Schools and Further Education Establishments* 1978

Conclusions

1 The scope of the Act is described in chapter 2

Appendix I

This appendix is divided into three parts, A, B and C.

Part A consists of the whole of the Health and Safety at Work etc Act 1974, excluding Part III (Building Regulations and amendment of the Building (Scotland) Act 1969).

Part B reproduces Schedule 15 of The Employment Protection Act 1975. (It is included because it makes minor changes to the HASAW Act 1974.)

Part C reproduces The Safety Representatives and Safety Committees Regulations 1977, these are reproduced in full so as to be available for use with chapter 10.

Health and Safety at Work etc. Act 1974

CHAPTER 37

ARRANGEMENT OF SECTIONS

PART I

HEALTH, SAFETY AND WELFARE IN CONNECTION WITH WORK, AND CONTROL OF DANGEROUS SUBSTANCES AND CERTAIN EMISSIONS INTO THE ATMOSPHERE

Enforcement

Miscellaneous and supplementary

PART II

THE EMPLOYMENT MEDICAL ADVISORY SERVICE

PART III

BUILDING REGULATIONS AND AMENDMENT OF BUILDING (SCOTLAND) ACT 1959

Section

PART IV

MISCELLANEOUS AND GENERAL

SCHEDULES:

Health and Safety at Work etc. Act 1974

1974 CHAPTER 37

An Act to make further provision for securing the health, safety and welfare of persons at work, for protecting others against risks to health or safety in connection with the activities of persons at work, for controlling the keeping and use and preventing the unlawful acquisition, possession and use of dangerous substances, and for controlling certain emissions into the atmosphere; to make further provision with respect to the employment medical advisory service; to amend the law relating to building regulations, and the Building (Scotland) Act 1959; and for connected purposes. [31st July 1974]

BE IT ENACTED by the Queen's most Excellent Majesty, by and with the advice and consent of the Lords Spiritual and Temporal, and Commons, in this present Parliament assembled, and by the authority of the same, as follows:—

PART I

HEALTH, SAFETY AND WELFARE IN CONNECTION WITH WORK, AND CONTROL OF DANGEROUS SUBSTANCES AND CERTAIN EMISSIONS INTO THE ATMOSPHERE

Preliminary

1.—(1) The provisions of this Part shall have effect with a view to— Preliminary.

(*a*) securing the health, safety and welfare of persons at work;

(*b*) protecting persons other than persons at work against risks to health or safety arising out of or in connection with the activities of persons at work;

PART I

(*c*) controlling the keeping and use of explosive or highly flammable or otherwise dangerous substances, and generally preventing the unlawful acquisition, possession and use of such substances; and

(*d*) controlling the emission into the atmosphere of noxious or offensive substances from premises of any class prescribed for the purposes of this paragraph.

(2) The provisions of this Part relating to the making of health and safety regulations and agricultural health and safety regulations and the preparation and approval of codes of practice shall in particular have effect with a view to enabling the enactments specified in the third column of Schedule 1 and the regulations, orders and other instruments in force under those enactments to be progressively replaced by a system of regulations and approved codes of practice operating in combination with the other provisions of this Part and designed to maintain or improve the standards of health, safety and welfare established by or under those enactments.

(3) For the purposes of this Part risks arising out of or in connection with the activities of persons at work shall be treated as including risks attributable to the manner of conducting an undertaking, the plant or substances used for the purposes of an undertaking and the condition of premises so used or any part of them.

(4) References in this Part to the general purposes of this Part are references to the purposes mentioned in subsection (1) above.

General duties

General duties of employers to their employees.

2.—(1) It shall be the duty of every employer to ensure, so far as is reasonably practicable, the health, safety and welfare at work of all his employees.

(2) Without prejudice to the generality of an employer's duty under the preceding subsection, the matters to which that duty extends include in particular—

(*a*) the provision and maintenance of plant and systems of work that are, so far as is reasonably practicable, safe and without risks to health;

(*b*) arrangements for ensuring, so far as is reasonably practicable, safety and absence of risks to health in connection with the use, handling, storage and transport of articles and substances;

(*c*) the provision of such information, instruction, training and supervision as is necessary to ensure, so far as is reasonably practicable, the health and safety at work of his employees;

(*d*) so far as is reasonably practicable as regards any place of work under the employer's control, the maintenance of it in a condition that is safe and without risks to health and the provision and maintenance of means of access to and egress from it that are safe and without such risks;

(*e*) the provision and maintenance of a working environment for his employees that is, so far as is reasonably practicable, safe, without risks to health, and adequate as regards facilities and arrangements for their welfare at work.

(3) Except in such cases as may be prescribed, it shall be the duty of every employer to prepare and as often as may be appropriate revise a written statement of his general policy with respect to the health and safety at work of his employees and the organisation and arrangements for the time being in force for carrying out that policy, and to bring the statement and any revision of it to the notice of all of his employees.

(4) Regulations made by the Secretary of State may provide for the appointment in prescribed cases by recognised trade unions (within the meaning of the regulations) of safety representatives from amongst the employees, and those representatives shall represent the employees in consultations with the employers under subsection (6) below and shall have such other functions as may be prescribed.

(5) Regulations made by the Secretary of State may provide for the election in prescribed cases by employees of safety representatives from amongst the employees, and those representatives shall represent the employees in consultations with the employers under subsection (6) below and may have such other functions as may be prescribed.

(6) It shall be the duty of every employer to consult any such representatives with a view to the making and maintenance of arrangements which will enable him and his employees to co-operate effectively in promoting and developing measures to ensure the health and safety at work of the employees, and in checking the effectiveness of such measures.

(7) In such cases as may be prescribed it shall be the duty of every employer, if requested to do so by the safety representatives mentioned in subsections (4) and (5) above, to establish, in accordance with regulations made by the Secretary of State, a safety committee having the function of keeping under review the measures taken to ensure the health and safety at work of his employees and such other functions as may be prescribed.

PART I

General duties of employers and self-employed to persons other than their employees.

3.—(1) It shall be the duty of every employer to conduct his undertaking in such a way as to ensure, so far as is reasonably practicable, that persons not in his employment who may be affected thereby are not thereby exposed to risks to their health or safety.

(2) It shall be the duty of every self-employed person to conduct his undertaking in such a way as to ensure, so far as is reasonably practicable, that he and other persons (not being his employees) who may be affected thereby are not thereby exposed to risks to their health or safety.

(3) In such cases as may be prescribed, it shall be the duty of every employer and every self-employed person, in the prescribed circumstances and in the prescribed manner, to give to persons (not being his employees) who may be affected by the way in which he conducts his undertaking the prescribed information about such aspects of the way in which he conducts his undertaking as might affect their health or safety.

General duties of persons concerned with premises to persons other than their employees.

4.—(1) This section has effect for imposing on persons duties in relation to those who—

(*a*) are not their employees ; but

(*b*) use non-domestic premises made available to them as a place of work or as a place where they may use plant or substances provided for their use there,

and applies to premises so made available and other non-domestic premises used in connection with them.

(2) It shall be the duty of each person who has, to any extent, control of premises to which this section applies or of the means of access thereto or egress therefrom or of any plant or substance in such premises to take such measures as it is reasonable for a person in his position to take to ensure, so far as is reasonably practicable, that the premises, all means of access thereto or egress therefrom available for use by persons using the premises, and any plant or substance in the premises or, as the case may be, provided for use there, is or are safe and without risks to health.

(3) Where a person has, by virtue of any contract or tenancy an obligation of any extent in relation to—

(*a*) the maintenance or repair of any premises to which this section applies or any means of access thereto or egress therefrom ; or

(*b*) the safety of or the absence of risks to health arising from plant or substances in any such premises ;

that person shall be treated, for the purposes of subsection (2) above, as being a person who has control of the matters to which his obligation extends.

(4) Any reference in this section to a person having control of any premises or matter is a reference to a person having control of the premises or matter in connection with the carrying on by him of a trade, business or other undertaking (whether for profit or not).

General duty of persons in control of certain premises in relation to harmful emissions into atmosphere.

5.—(1) It shall be the duty of the person having control of any premises of a class prescribed for the purposes of section 1(1)(*d*) to us the best practicable means for preventing the emission into the atmosphere from the premises of noxious or offensive substances and for rendering harmless and inoffensive such substances as may be so emitted.

(2) The reference in subsection (1) above to the means to be used for the purposes there mentioned includes a reference to the manner in which the plant provided for those purposes is used and to the supervision of any operation involving the emission of the substances to which that subsection applies.

(3) Any substance or a substance of any description prescribed for the purposes of subsection (1) above as noxious or offensive shall be a noxious or, as the case may be, an offensive substance for those purposes whether or not it would be so apart from this subsection.

(4) Any reference in this section to a person having control of any premises is a reference to a person having control of the premises in connection with the carrying on by him of a trade, business or other undertaking (whether for profit or not) and any duty imposed on any such person by this section shall extend only to matters within his control.

General duties of manufacturers etc. as regards articles and substances for use at work.

6.—(1) It shall be the duty of any person who designs, manufactures, imports or supplies any article for use at work—

(*a*) to ensure, so far as is reasonably practicable, that the article is so designed and constructed as to be safe and without risks to health when properly used;

(*b*) to carry out or arrange for the carrying out of such testing and examination as may be necessary for the performance of the duty imposed on him by the preceding paragraph;

(*c*) to take such steps as are necessary to secure that there will be available in connection with the use of the article at work adequate information about the use for which it is designed and has been tested, and about any conditions necessary to ensure that, when put to that use, it will be safe and without risks to health.

(2) It shall be the duty of any person who undertakes the design or manufacture of any article for use at work to carry out

PART I

or arrange for the carrying out of any necessary research with a view to the discovery and, so far as is reasonably practicable, the elimination or minimisation of any risks to health or safety to which the design or article may give rise.

(3) It shall be the duty of any person who erects or installs any article for use at work in any premises where that article is to be used by persons at work to ensure, so far as is reasonably practicable, that nothing about the way in which it is erected or installed makes it unsafe or a risk to health when properly used.

(4) It shall be the duty of any person who manufactures, imports or supplies any substance for use at work—

(*a*) to ensure, so far as is reasonably practicable, that the substance is safe and without risks to health when properly used ;

(*b*) to carry out or arrange for the carrying out of such testing and examination as may be necessary for the performance of the duty imposed on him by the preceding paragraph ;

(*c*) to take such steps as are necessary to secure that there will be available in connection with the use of the substance at work adequate information about the results of any relevant tests which have been carried out on or in connection with the substance and about any conditions necessary to ensure that it will be safe and without risks to health when properly used.

(5) It shall be the duty of any person who undertakes the manufacture of any substance for use at work to carry out or arrange for the carrying out of any necessary research with a view to the discovery and, so far as is reasonably practicable, the elimination or minimisation of any risks to health or safety to which the substance may give rise.

(6) Nothing in the preceding provisions of this section shall be taken to require a person to repeat any testing, examination or research which has been carried out otherwise than by him or at his instance, in so far as it is reasonable for him to rely on the results thereof for the purposes of those provisions.

(7) Any duty imposed on any person by any of the preceding provisions of this section shall extend only to things done in the course of a trade, business or other undertaking carried on by him (whether for profit or not) and to matters within his control.

(8) Where a person designs, manufactures, imports or supplies an article for or to another on the basis of a written undertaking by that other to take specified steps sufficient to ensure, so far as is reasonably practicable, that the article will be safe and

without risks to health when properly used, the undertaking shall have the effect of relieving the first-mentioned person from the duty imposed by subsection (1)(*a*) above to such extent as is reasonable having regard to the terms of the undertaking.

(9) Where a person (" the ostensible supplier ") supplies any article for use at work or substance for use at work to another (" the customer ") under a hire-purchase agreement, conditional sale agreement or credit-sale agreement, and the ostensible supplier—

(*a*) carries on the business of financing the acquisition of goods by others by means of such agreements; and

(*b*) in the course of that business acquired his interest in the article or substance supplied to the customer as a means of financing its acquisition by the customer from a third person (" the effective supplier "),

the effective supplier and not the ostensible supplier shall be treated for the purposes of this section as supplying the article or substance to the customer, and any duty imposed by the preceding provisions of this section on suppliers shall accordingly fall on the effective supplier and not on the ostensible supplier.

(10) For the purposes of this section an article or substance is not to be regarded as properly used where it is used without regard to any relevant information or advice relating to its use which has been made available by a person by whom it was designed, manufactured, imported or supplied.

General duties of employees at work.

7. It shall be the duty of every employee while at work—

(*a*) to take reasonable care for the health and safety of himself and of other persons who may be affected by his acts or omissions at work; and

(*b*) as regards any duty or requirement imposed on his employer or any other person by or under any of the relevant statutory provisions, to co-operate with him so far as is necessary to enable that duty or requirement to be performed or complied with.

Duty not to interfere with or misuse things provided pursuant to certain provisions.

8. No person shall intentionally or recklessly interfere with or misuse anything provided in the interests of health, safety or welfare in pursuance of any of the relevant statutory provisions.

Duty not to charge employees for things done or provided pursuant to certain specfic requirements.

9. No employer shall levy or permit to be levied on any employee of his any charge in respect of anything done or provided in pursuance of any specific requirement of the relevant statutory provisions.

PART I

The Health and Safety Commission and the Health and Safety Executive

Establishment of the Commission and the Executive.

10.—(1) There shall be two bodies corporate to be called the Health and Safety Commission and the Health and Safety Executive which shall be constituted in accordance with the following provisions of this section.

(2) The Health and Safety Commission (hereafter in this Act referred to as " the Commission ") shall consist of a chairman appointed by the Secretary of State and not less than six nor more than nine other members appointed by the Secretary of State in accordance with subsection (3) below.

(3) Before appointing the members of the Commission (other than the chairman) the Secretary of State shall—

(*a*) as to three of them, consult such organisations representing employers as he considers appropriate;

(*b*) as to three others, consult such organisations representing employees as he considers appropriate; and

(*c*) as to any other members he may appoint, consult such organisations representing local authorities and such other organisations, including professional bodies, the activities of whose members are concerned with matters relating to any of the general purposes of this Part, as he considers appropriate.

(4) The Secretary of State may appoint one of the members to be deputy chairman of the Commission.

(5) The Health and Safety Executive (hereafter in this Act referred to as " the Executive ") shall consist of three persons of whom one shall be appointed by the Commission with the approval of the Secretary of State to be the director of the Executive and the others shall be appointed by the Commission with the like approval after consultation with the said director.

(6) The provisions of Schedule 2 shall have effect with respect to the Commission and the Executive.

(7) The functions of the Commission and of the Executive, and of their officers and servants, shall be performed on behalf of the Crown.

General functions of the Commission and the Executive.

11.—(1) In addition to the other functions conferred on the Commission by virtue of this Act, but subject to subsection (3) below, it shall be the general duty of the Commission to do such things and make such arrangements as it considers appropriate for the general purposes of this Part except as regards matters relating exclusively to agricultural operations.

(2) It shall be the duty of the Commission, except as aforesaid— PART I

(*a*) to assist and encourage persons concerned with matters relevant to any of the general purposes of this Part to further those purposes ;

(*b*) to make such arrangements as it considers appropriate for the carrying out of research, the publication of the results of research and the provision of training and information in connection with those purposes, and to encourage research and the provision of training and information in that connection by others ;

(*c*) to make such arrangements as it considers appropriate for securing that government departments, employers, employees, organisations representing employers and employees respectively, and other persons concerned with matters relevant to any of those purposes are provided with an information and advisory service and are kept informed of, and adequately advised on, such matters ;

(*d*) to submit from time to time to the authority having power to make regulations under any of the relevant statutory provisions such proposals as the Commission considers appropriate for the making of regulations under that power.

(3) It shall be the duty of the Commission—

(*a*) to submit to the Secretary of State from time to time particulars of what it proposes to do for the purpose of performing its functions ; and

(*b*) subject to the following paragraph, to ensure that its activities are in accordance with proposals approved by the Secretary of State ; and

(*c*) to give effect to any directions given to it by the Secretary of State.

(4) In addition to any other functions conferred on the Executive by virtue of this Part, it shall be the duty of the Executive—

(*a*) to exercise on behalf of the Commission such of the Commission's functions as the Commission directs it to exercise ; and

(*b*) to give effect to any directions given to it by the Commission otherwise than in pursuance of paragraph (*a*) above ;

but, except for the purpose of giving effect to directions given to the Commission by the Secretary of State, the Commission shall not give to the Executive any directions as to the enforcement of any of the relevant statutory provisions in a particular case.

PART I

(5) Without prejudice to subsection (2) above, it shall be the duty of the Executive, if so requested by a Minister of the Crown—

(*a*) to provide him with information about the activities of the Executive in connection with any matter with which he is concerned; and

(*b*) to provide him with advice on any matter with which he is concerned on which relevant expert advice is obtainable from any of the officers or servants of the Executive but which is not relevant to any of the general purposes of this Part.

(6) The Commission and the Executive shall, subject to any directions given to it in pursuance of this Part, have power to do anything (except borrow money) which is calculated to facilitate, or is conducive or incidental to, the performance of any function of the Commission or, as the case may be, the Executive (including a function conferred on it by virtue of this subsection).

Control of the Commission by the Secretary of State.

12. The Secretary of State may—

(*a*) approve with or without modifications, any proposals submitted to him in pursuance of section 11(3)(*a*);

(*b*) give to the Commission at any time such directions as he thinks fit with respect to its functions (including directions modifying its functions, but not directions conferring on it functions other than any of which it was deprived by previous directions given by virtue of this paragraph), and any directions which it appears to him requisite or expedient to give in the interests of the safety of the State.

Other powers of the Commission.

13.—(1) The Commission shall have power—

(*a*) to make agreements with any government department or other person for that department or person to perform on behalf of the Commission or the Executive (with or without payment) any of the functions of the Commission or, as the case may be, of the Executive;

(*b*) subject to subsection (2) below, to make agreements with any Minister of the Crown, government department or other public authority for the Commission to perform on behalf of that Minister, department or authority (with or without payment) functions exercisable by the Minister, department or authority (including, in the case of a Minister, functions not conferred by an enactment), being functions which in the opinion of the Secretary of State can appropriately be performed by the Commission in connection with any of the Commission's functions;

(*c*) to provide (with or without payment) services or facilities required otherwise than for the general purposes of this Part in so far as they are required by any government department or other public authority in connection with the exercise by that department or authority of any of its functions;

(*d*) to appoint persons or committees of persons to provide the Commission with advice in connection with any of its functions and (without prejudice to the generality of the following paragraph) to pay to persons so appointed such remuneration as the Secretary of State may with the approval of the Minister for the Civil Service determine;

(*e*) in connection with any of the functions of the Commission, to pay to any person such travelling and subsistence allowances and such compensation for loss of remunerative time as the Secretary of State may with the approval of the Minister for the Civil Service determine;

(*f*) to carry out or arrange for or make payments in respect of research into any matter connected with any of the Commission's functions, and to disseminate or arrange for or make payments in respect of the dissemination of information derived from such research;

(*g*) to include, in any arrangements made by the Commission for the provision of facilities or services by it or on its behalf, provision for the making of payments to the Commission or any person acting on its behalf by other parties to the arrangements and by persons who use those facilities or services.

(2) Nothing in subsection (1)(*b*) shall authorise the Commission to perform any function of a Minister, department or authority which consists of a power to make regulations or other instruments of a legislative character.

Power of the Commission to direct investigations and inquiries.

14.—(1) This section applies to the following matters, that is to say any accident, occurrence, situation or other matter whatsoever which the Commission thinks it necessary or expedient to investigate for any of the general purposes of this Part or with a view to the making of regulations for those purposes; and for the purposes of this subsection it is immaterial whether the Executive is or is not responsible for securing the enforcement of such (if any) of the relevant statutory provisions as relate to the matter in question.

(2) The Commission may at any time—

(*a*) direct the Executive or authorise any other person to investigate and make a special report on any matter to which this section applies; or

PART I

(*b*) with the consent of the Secretary of State direct an inquiry to be held into any such matter;

but shall not do so in any particular case that appears to the Commission to involve only matters relating exclusively to agricultural operation.

(3) Any inquiry held by virtue of subsection (2)(*b*) above shall be held in accordance with regulations made for the purposes of this subsection by the Secretary of State, and shall be held in public except where or to the extent that the regulations provide otherwise.

(4) Regulations made for the purposes of subsection (3) above may in particular include provision—

(*a*) conferring on the person holding any such inquiry, and any person assisting him in the inquiry, powers of entry and inspection;

(*b*) conferring on any such person powers of summoning witnesses to give evidence or produce documents and power to take evidence on oath and administer oaths or require the making of declarations;

(*c*) requiring any such inquiry to be held otherwise than in public where or to the extent that a Minister of the Crown so directs.

(5) In the case of a special report made by virtue of subsection (2)(*a*) above or a report made by the person holding an inquiry held by virtue of subsection (2)(*b*) above, the Commission may cause the report, or so much of it as the Commission thinks fit, to be made public at such time and in such manner as the Commission thinks fit.

(6) The Commission—

(*a*) in the case of an investigation and special report made by virtue of subsection (2)(*a*) above (otherwise than by an officer or servant of the Executive), may pay to the person making it such remuneration and expenses as the Secretary of State may, with the approval of the Minister for the Civil Service, determine;

(*b*) in the case of an inquiry held by virtue of subsection (2)(*b*) above, may pay to the person holding it and to any assessor appointed to assist him such remuneration and expenses, and to persons attending the inquiry as witnesses such expenses, as the Secretary of State may, with the like approval, determine; and

(*c*) may, to such extent as the Secretary of State may determine, defray the other costs, if any, of any such investigation and special report or inquiry.

PART I

(7) Where an inquiry is directed to be held by virtue of subsection (2)(*b*) above into any matter to which this section applies arising in Scotland, being a matter which causes the death of any person, no inquiry with regard to that death shall, unless the Lord Advocate otherwise directs, be held in pursuance of the Fatal Accidents Inquiry (Scotland) Act 1895. 1895 c. 36.

Health and safety regulations and approved codes of practice

Health and safety regulations.

15.—(1) Subject to the provisions of section 50, the Secretary of State shall have power to make regulations under this section (in this part referred to as " health and safety regulations ") for any of the general purposes of this Part except as regards matters relating exclusively to agricultural operations.

(2) Without prejudice to the generality of the preceding subsection, health and safety regulations may for any of the general purposes of this Part make provision for any of the purposes mentioned in Schedule 3.

(3) Health and safety regulations—

(*a*) may repeal or modify any of the existing statutory provisions ;

(*b*) may exclude or modify in relation to any specified class of case any of the provisions of sections 2 to 9 or any of the existing statutory provisions ;

(*c*) may make a specified authority or class of authorities responsible, to such extent as may be specified, for the enforcement of any of the relevant statutory provisions.

(4) Health and safety regulations—

(*a*) may impose requirements by reference to the approval of the Commission or any other specified body or person ;

(*b*) may provide for references in the regulations to any specified document to operate as references to that document as revised or re-issued from time to time.

(5) Health and safety regulations—

(*a*) may provide (either unconditionally or subject to conditions, and with or without limit of time) for exemptions from any requirement or prohibition imposed by or under any of the relevant statutory provisions ;

(*b*) may enable exemptions from any requirement or prohibition imposed by or under any of the relevant statutory provisions to be granted (either unconditionally or subject to conditions, and with or without limit of time) by any specified person or by any person authorised in that behalf by a specified authority.

(6) Health and safety regulations—

(*a*) may specify the persons or classes of persons who, in the event of a contravention of a requirement or prohibition imposed by or under the regulations, are to be guilty of an offence, whether in addition to or to the exclusion of other persons or classes of persons ;

(*b*) may provide for any specified defence to be available in proceedings for any offence under the relevant statutory provisions either generally or in specified circumstances ;

(*c*) may exclude proceedings on indictment in relation to offences consisting of a contravention of a requirement or prohibition imposed by or under any of the existing statutory provisions, sections 2 to 9 or health and safety regulations ;

(*d*) may restrict the punishments which can be imposed in respect of any such offence as is mentioned in paragraph (*c*) above.

(7) Without prejudice to section 35, health and safety regulations may make provision for enabling offences under any of the relevant statutory provisions to be treated as having been committed at any specified place for the purpose of bringing any such offence within the field of responsibility of any enforcing authority or conferring jurisdiction on any court to entertain proceedings for any such offence.

(8) Health and safety regulations may take the form of regulations applying to particular circumstances only or to a particular case only (for example, regulations applying to particular premises only).

(9) If an Order in Council is made under section 84(3) providing that this section shall apply to or in relation to persons, premises or work outside Great Britain then, notwithstanding the Order, health and safety regulations shall not apply to or in relation to aircraft in flight, vessels, hovercraft or offshore installations outside Great Britain or persons at work outside Great Britain in connection with submarine cables or submarine pipelines except in so far as the regulations expressly so provide.

(10) In this section " specified " means specified in health and safety regulations.

PART I

Approval of codes of practice by the Commission.

16.—(1) For the purpose of providing practical guidance with respect to the requirements of any provision of sections 2 to 7 or of health and safety regulations or of any of the existing statutory provisions, the Commission may, subject to the following subsection and except as regards matters relating exclusively to agricultural operations—

(*a*) approve and issue such codes of practice (whether prepared by it or not) as in its opinion are suitable for that purpose;

(*b*) approve such codes of practice issued or proposed to be issued otherwise than by the Commission as in its opinion are suitable for that purpose.

(2) The Commission shall not approve a code of practice under subsection (1) above without the consent of the Secretary of State, and shall, before seeking his consent, consult—

(*a*) any government department or other body that appears to the Commission to be appropriate (and, in particular, in the case of a code relating to electromagnetic radiations, the National Radiological Protection Board); and

(*b*) such government departments and other bodies, if any, as in relation to any matter dealt with in the code, the Commission is required to consult under this section by virtue of directions given to it by the Secretary of State.

(3) Where a code of practice is approved by the Commission under subsection (1) above, the Commission shall issue a notice in writing—

(*a*) identifying the code in question and stating the date on which its approval by the Commission is to take effect; and

(*b*) specifying for which of the provisions mentioned in subsection (1) above the code is approved.

(4) The Commission may—

(*a*) from time to time revise the whole or any part of any code of practice prepared by it in pursuance of this section;

(*b*) approve any revision or proposed revision of the whole or any part of any code of practice for the time being approved under this section;

and the provisions of subsections (2) and (3) above shall, with the necessary modifications, apply in relation to the approval of any revision under this subsection as they apply in relation to the approval of a code of practice under subsection (1) above.

(5) The Commission may at any time with the consent of the Secretary of State withdraw its approval from any code of practice approved under this section, but before seeking his consent shall consult the same government departments and other bodies as it would be required to consult under subsection (2) above if it were proposing to approve the code.

(6) Where under the preceding subsection the Commission withdraws its approval from a code of practice approved under this section, the Commission shall issue a notice in writing identifying the code in question and stating the date on which its approval of it is to cease to have effect.

(7) References in this Part to an approved code of practice are references to that code as it has effect for the time being by virtue of any revision of the whole or any part of it approved under this section.

(8) The power of the Commission under subsection (1)(*b*) above to approve a code of practice issued or proposed to be issued otherwise than by the Commission shall include power to approve a part of such a code of practice; and accordingly in this Part "code of practice" may be read as including a part of such a code of practice.

Use of approved codes of practice in criminal proceedings.

17.—(1) A failure on the part of any person to observe any provision of an approved code of practice shall not of itself render him liable to any civil or criminal proceedings; but where in any criminal proceedings a party is alleged to have committed an offence by reason of a contravention of any requirement or prohibition imposed by or under any such provision as is mentioned in section 16(1) being a provision for which there was an approved code of practice at the time of the alleged contravention, the following subsection shall have effect with respect to that code in relation to those proceedings.

(2) Any provision of the code of practice which appears to the court to be relevant to the requirement or prohibition alleged to have been contravened shall be admissible in evidence in the proceedings; and if it is proved that there was at any material time a failure to observe any provision of the code which appears to the court to be relevant to any matter which it is necessary for the prosecution to prove in order to establish a contravention of that requirement or prohibition, that matter shall be taken as proved unless the court is satisfied that the requirement or prohibition was in respect of that matter complied with otherwise than by way of observance of that provision of the code.

(3) In any criminal proceedings—

(*a*) a document purporting to be a notice issued by the Commission under section 16 shall be taken to be such a notice unless the contrary is proved; and (*b*) a code of practice which appears to the court to be the subject of such a notice shall be taken to be the subject of that notice unless the contrary is proved.

Enforcement

Authorities responsible for enforcement of the relevant statutory provisions.

18.—(1) It shall be the duty of the Executive to make adequate arrangements for the enforcement of the relevant statutory provisions except to the extent that some other authority or class of authorities is by any of those provisions or by regulations under subsection (2) below made responsible for their enforcement.

(2) The Secretary of State may by regulations—

(*a*) make local authorities responsible for the enforcement of the relevant statutory provisions to such extent as may be prescribed;

(*b*) make provision for enabling responsibility for enforcing any of the relevant statutory provisions to be, to such extent as may be determined under the regulations—

(i) transferred from the Executive to local authorities or from local authorities to the Executive; or

(ii) assigned to the Executive or to local authorities for the purpose of removing any uncertainty as to what are by virtue of this subsection their respective responsibilities for the enforcement of those provisions;

and any regulations made in pursuance of paragraph (*b*) above shall include provision for securing that any transfer or assignment effected under the regulations is brought to the notice of persons affected by it.

(3) Any provision made by regulations under the preceding subsection shall have effect subject to any provision made by health and safety regulations or agricultural health and safety regulations in pursuance of section 15(3)(*c*).

(4) It shall be the duty of every local authority—

(*a*) to make adequate arrangements for the enforcement within their area of the relevant statutory provisions to the extent that they are by any of those provisions or by regulations under subsection (2) above made responsible for their enforcement; and

(*b*) to perform the duty imposed on them by the preceding paragraph and any other functions conferred on them by any of the relevant statutory provisions in accordance with such guidance as the Commission may give them.

PART I

(5) Where any authority other than the appropriate Agriculture Minister, the Executive or a local authority is by any of the relevant statutory provisions or by regulations under subsection (2) above made responsible for the enforcement of any of those provisions to any extent, it shall be the duty of that authority—

(*a*) to make adequate arrangements for the enforcement of those provisions to that extent; and

(*b*) to perform the duty imposed on the authority by the preceding paragraph and any other functions conferred on the authority by any of the relevant statutory provisions in accordance with such guidance as the Commission may give to the authority.

(6) Nothing in the provisions of this Act or of any regulations made thereunder charging any person in Scotland with the enforcement of any of the relevant statutory provisions shall be construed as authorising that person to institute proceedings for any offence.

(7) In this Part—

(*a*) " enforcing authority " means the Executive or any other authority which is by any of the relevant statutory provisions or by regulations under subsection (2) above made responsible for the enforcement of any of those provisions to any extent; and

(*b*) any reference to an enforcing authority's field of responsibility is a reference to the field over which that authority's responsibility for the enforcement of those provisions extends for the time being;

but where by virtue of paragraph (*a*) of section 13(1) the performance of any function of the Commission or the Executive is delegated to a government department or person, references to the Commission or the Executive (or to an enforcing authority where that authority is the Executive) in any provision of this Part which relates to that function shall, so far as may be necessary to give effect to any agreement under that paragraph, be construed as references to that department or person; and accordingly any reference to the field of responsibility of an enforcing authority shall be construed as a reference to the field over which that department or person for the time being performs such a function.

Appointment of inspectors.

19.—(1) Every enforcing authority may appoint as inspectors (under whatever title it may from time to time determine) such persons having suitable qualifications as it thinks necessary for carrying into effect the relevant statutory provisions within its field of responsibility, and may terminate any appointment made under this section.

PART I

(2) Every appointment of a person as an inspector under this section shall be made by an instrument in writing specifying which of the powers conferred on inspectors by the relevant statutory provisions are to be exercisable by the person appointed; and an inspector shall in right of his appointment under this section—

(*a*) be entitled to exercise only such of those powers as are so specified; and

(*b*) be entitled to exercise the powers so specified only within the field of responsibility of the authority which appointed him.

(3) So much of an inspector's instrument of appointment as specifies the powers which he is entitled to exercise may be varied by the enforcing authority which appointed him.

(4) An inspector shall, if so required when exercising or seeking to exercise any power conferred on him by any of the relevant statutory provisions, produce his instrument of appointment or a duly authenticated copy thereof.

Powers of inspectors.

20.—(1) Subject to the provisions of section 19 and this section, an inspector may, for the purpose of carrying into effect any of the relevant statutory provisions within the field of responsibility of the enforcing authority which appointed him, exercise the powers set out in subsection (2) below.

(2) The powers of an inspector referred to in the preceding subsection are the following, namely—

(*a*) at any reasonable time (or, in a situation which in his opinion is or may be dangerous, at any time) to enter any premises which he has reason to believe it is necessary for him to enter for the purpose mentioned in subsection (1) above;

(*b*) to take with him a constable if he has reasonable cause to apprehend any serious obstruction in the execution of his duty;

(*c*) without prejudice to the preceding paragraph, on entering any premises by virtue of paragraph (*a*) above to take with him—

(i) any other person duly authorised by his (the inspector's) enforcing authority; and

(ii) any equipment or materials required for any purpose for which the power of entry is being exercised;

(*d*) to make such examination and investigation as may in any circumstances be necessary for the purpose mentioned in subsection (1) above;

PART I

(*e*) as regards any premises which he has power to enter, to direct that those premises or any part of them, or anything therein, shall be left undisturbed (whether generally or in particular respects) for so long as is reasonably necessary for the purpose of any examination or investigation under paragraph (*d*) above ;

(*f*) to take such measurements and photographs and make such recordings as he considers necessary for the purpose of any examination or investigation under paragraph (*d*) above ;

(*g*) to take samples of any articles or substances found in any premises which he has power to enter, and of the atmosphere in or in the vicinity of any such premises ;

(*h*) in the case of any article or substance found in any premises which he has power to enter, being an article or substance which appears to him to have caused or to be likely to cause danger to health or safety, to cause it to be dismantled or subjected to any process or test (but not so as to damage or destroy it unless this is in the circumstances necessary for the purpose mentioned in subsection (1) above) ;

(*i*) in the case of any such article or substance as is mentioned in the preceding paragraph, to take possession of it and detain it for so long as is necessary for all or any of the following purposes, namely—

(i) to examine it and do to it anything which he has power to do under that paragraph ;

(ii) to ensure that it is not tampered with before his examination of it is completed ;

(iii) to ensure that it is available for use as evidence in any proceedings for an offence under any of the relevant statutory provisions or any proceedings relating to a notice under section 21 or 22 ;

(*j*) to require any person whom he has reasonable cause to believe to be able to give any information relevant to any examination or investigation under paragraph (*d*) above to answer (in the absence of persons other than a person nominated by him to be present and any persons whom the inspector may allow to be present) such questions as the inspector thinks fit to ask and to sign a declaration of the truth of his answers ;

(*k*) to require the production of, inspect, and take copies of or of any entry in—

(i) any books or documents which by virtue of any of the relevant statutory provisions are required to be kept ; and

(ii) any other books or documents which it is necessary for him to see for the purposes of any examination or investigation under paragraph (*d*) above;

(*l*) to require any person to afford him such facilities and assistance with respect to any matter or things within that person's control or in relation to which that person has responsibilities as are necessary to enable the inspector to exercise any of the powers conferred on him by this section;

(*m*) any other power which is necessary for the purpose mentioned in subsection (1) above.

(3) The Secretary of State may by regulations make provision as to the procedure to be followed in connection with the taking of samples under subsection (2)(*g*) above (including provision as to the way in which samples that have been so taken are to be dealt with).

(4) Where an inspector proposes to exercise the power conferred by subsection (2)(*h*) above in the case of an article or substance found in any premises, he shall, if so requested by a person who at the time is present in and has responsibilities in relation to those premises, cause anything which is to be done by virtue of that power to be done in the presence of that person unless the inspector considers that its being done in that person's presence would be prejudicial to the safety of the State.

(5) Before exercising the power conferred by subsection (2)(*h*) above in the case of any article or substance, an inspector shall consult such persons as appear to him appropriate for the purpose of ascertaining what dangers, if any, there may be in doing anything which he proposes to do under that power.

(6) Where under the power conferred by subsection (2)(*i*) above an inspector takes possession of any article or substance found in any premises, he shall leave there, either with a responsible person or, if that is impracticable, fixed in a conspicuous position, a notice giving particulars of that article or substance sufficient to identify it and stating that he has taken possession of it under that power; and before taking possession of any such substance under that power an inspector shall, if it is practicable for him to do so, take a sample thereof and give to a responsible person at the premises a portion of the sample marked in a manner sufficient to identify it.

(7) No answer given by a person in pursuance of a requirement imposed under subsection (2)(*j*) above shall be admissible in evidence against that person or the husband or wife of that person in any proceedings.

(8) Nothing in this section shall be taken to compel the production by any person of a document of which he would on grounds of legal professional privilege be entitled to withhold production on an order for discovery in an action in the High Court or, as the case may be, on an order for the production of documents in an action in the Court of Session.

Improvement notices.

21. If an inspector is of the opinion that a person—

(*a*) is contravening one or more of the relevant statutory provisions ; or

(*b*) has contravened one or more of those provisions in circumstances that make it likely that the contravention will continue or be repeated,

he may serve on him a notice (in this Part referred to as " an improvement notice ") stating that he is of that opinion, specifying the provision or provisions as to which he is of that opinion, giving particulars of the reasons why he is of that opinion, and requiring that person to remedy the contravention or, as the case may be, the matters occasioning it within such period (ending not earlier than the period within which an appeal against the notice can be brought under section 24) as may be specified in the notice.

Prohibition notices.

22.—(1) This section applies to any activities which are being or are about to be carried on by or under the control of any person, being activities to or in relation to which any of the relevant statutory provisions apply or will, if the activities are so carried on, apply.

(2) If as regards any activities to which this section applies an inspector is of the opinion that, as carried on or about to be carried on by or under the control of the person in question, the activities involve or, as the case may be, will involve a risk of serious personal injury, the inspector may serve on that person a notice (in this Part referred to as " a prohibition notice ").

(3) A prohibition notice shall—

(*a*) state that the inspector is of the said opinion ;

(*b*) specify the matters which in his opinion give or, as the case may be, will give rise to the said risk ;

(*c*) where in his opinion any of those matters involves or, as the case may be, will involve a contravention of any of the relevant statutory provisions, state that he is of that opinion, specify the provision or provisions as to which he is of that opinion, and give particulars of the reasons why he is of that opinion ; and

(*d*) direct that the activities to which the notice relates shall not be carried on by or under the control of the person on whom the notice is served unless the matters

PART I

specified in the notice in pursuance of paragraph (*b*) above and any associated contraventions of provisions so specified in pursuance of paragraph (*c*) above have been remedied.

(4) A direction given in pursuance of subsection (3)(*d*) above shall take immediate effect if the inspector is of the opinion, and states it, that the risk of serious personal injury is or, as the case may be, will be imminent, and shall have effect at the end of a period specified in the notice in any other case.

Provisions supplementary to ss. 21 and 22.

23.—(1) In this section " a notice " means an improvement notice or a prohibition notice.

(2) A notice may (but need not) include directions as to the measures to be taken to remedy any contravention or matter to which the notice relates ; and any such directions—

(*a*) may be framed to any extent by reference to any approved code of practice ; and

(*b*) may be framed so as to afford the person on whom the notice is served a choice between different ways of remedying the contravention or matter.

(3) Where any of the relevant statutory provisions applies to a building or any matter connected with a building and an inspector proposes to serve an improvement notice relating to a contravention of that provision in connection with that building or matter, the notice shall not direct any measures to be taken to remedy the contravention of that provision which are more onerous than those necessary to secure conformity with the requirements of any building regulations for the time being in force to which that building or matter would be required to conform if the relevant building were being newly erected unless the provision in question imposes specific requirements more onerous than the requirements of any such building regulations to which the building or matter would be required to conform as aforesaid.

In this subsection " the relevant building ", in the case of a building, means that building, and, in the case of a matter connected with a building, means the building with which the matter is connected.

(4) Before an inspector serves in connection with any premises used or about to be used as a place of work a notice requiring or likely to lead to the taking of measures affecting the means of escape in case of fire with which the premises are or ought to be provided, he shall consult the fire authority.

In this subsection " fire authority " has the meaning assigned by section 43(1) of the Fire Precautions Act 1971. 1971 c. 40.

PART I

(5) Where an improvement notice or a prohibition notice which is not to take immediate effect has been served—

(*a*) the notice may be withdrawn by an inspector at any time before the end of the period specified therein in pursuance of section 21 or section 22(4) as the case may be; and

(*b*) the period so specified may be extended or further extended by an inspector at any time when an appeal against the notice is not pending.

(6) In the application of this section to Scotland—

(*a*) in subsection (3) for the words from " with the requirements " to " aforesaid " there shall be substituted the words—

" (*a*) to any provisions of the building standards regulations to which that building or matter would be required to conform if the relevant building were being newly erected ; or

1959 c. 24.

(*b*) where the sheriff, on an appeal to him under section 16 of the Building (Scotland) Act 1959—

(i) against an order under section 10 of that Act requiring the execution of operations necessary to make the building or matter conform to the building standards regulations, or

(ii) against an order under section 11 of that Act requiring the building or matter to conform to a provision of such regulations,

has varied the order, to any provisions of the building standards regulations referred to in paragraph (*a*) above as affected by the order as so varied,

unless the relevant statutory provision imposes specific requirements more onerous than the requirements of any provisions of building standards regulations as aforesaid or, as the case may be, than the requirements of the order as varied by the sheriff." ;

(*b*) after subsection (5) there shall be inserted the following subsection—

" (5A) In subsection (3) above ' building standards regulations ' has the same meaning as in section 3 of the Building (Scotland) Act 1959.".

Appeal against improvement or prohibition notice.

24.—(1) In this section " a notice " means an improvement notice or a prohibition notice.

(2) A person on whom a notice is served may within such period from the date of its service as may be prescribed appeal to an industrial tribunal; and on such an appeal the tribunal

may either cancel or affirm the notice and, if it affirms it, may be so either in its original form or with such modifications as the tribunal may in the circumstances think fit.

(3) Where an appeal under this section is brought against a notice within the period allowed under the preceding subsection, then—

(*a*) in the case of an improvement notice, the bringing of the appeal shall have the effect of suspending the operation of the notice until the appeal is finally disposed of or, if the appeal is withdrawn, until the withdrawal of the appeal;

(*b*) in the case of a prohibition notice, the bringing of the appeal shall have the like effect if, but only if, on the application of the appellant the tribunal so directs (and then only from the giving of the direction).

(4) One or more assessors may be appointed for the purposes of any proceedings brought before an industrial tribunal under this section.

Power to deal with cause of imminent danger.

25.—(1) Where, in the case of any article or substance found by him in any premises which he has power to enter, an inspector has reasonable cause to believe that, in the circumstances in which he finds it, the article or substance is a cause of imminent danger of serious personal injury, he may seize it and cause it to be rendered harmless (whether by destruction or otherwise).

(2) Before there is rendered harmless under this section—

(*a*) any article that forms part of a batch of similar articles; or

(*b*) any substance,

the inspector shall, if it is practicable for him to do so, take a sample thereof and give to a responsible person at the premises where the article or substance was found by him a portion of the sample marked in a manner sufficient to identify it.

(3) As soon as may be after any article or substance has been seized and rendered harmless under this section, the inspector shall prepare and sign a written report giving particulars of the circumstances in which the article or substance was seized and so dealt with by him, and shall—

(*a*) give a signed copy of the report to a responsible person at the premises where the article or substance was found by him; and

PART I

(*b*) unless that person is the owner of the article or substance, also serve a signed copy of the report on the owner;

and if, where paragraph (*b*) above applies, the inspector cannot after reasonable enquiry ascertain the name or address of the owner, the copy may be served on him by giving it to the person to whom a copy was given under the preceding paragraph.

Power of enforcing authorities to indemnify their inspectors.

26. Where an action has been brought against an inspector in respect of an act done in the execution or purported execution of any of the relevant statutory provisions and the circumstances are such that he is not legally entitled to require the enforcing authority which appointed him to indemnify him, that authority may, nevertheless, indemnify him against the whole or part of any damages and costs or expenses which he may have been ordered to pay or may have incurred, if the authority is satisfied that he honestly believed that the act complained of was within his powers and that his duty as an inspector required or entitled him to do it.

Obtaining and disclosure of information

Obtaining of information by the Commission, the Executive, enforcing authorities etc.

27.—(1) For the purpose of obtaining—

(*a*) any information which the Commission needs for the discharge of its functions; or

(*b*) any information which an enforcing authority needs for the discharge of the authority's functions,

the Commission may, with the consent of the Secretary of State, serve on any person a notice requiring that person to furnish to the Commission or, as the case may be, to the enforcing authority in question such information about such matters as may be specified in the notice, and to do so in such form and manner and within such time as may be so specified.

In this subsection " consent " includes a general consent extending to cases of any stated description.

1947 c. 39.

(2) Nothing in section 9 of the Statistics of Trade Act 1947 (which restricts the disclosure of information obtained under that Act) shall prevent or penalise—

(*a*) the disclosure by a Minister of the Crown to the Commission or the Executive of information obtained under that Act about any undertaking within the meaning of that Act, being information consisting of the names and address of the persons carrying on the undertaking, the nature of the undertaking's activities, the numbers of persons of different descriptions who work in the undertaking, the addresses or places where activities of the undertaking are or were carried on, the nature of the activities carried on there, or the numbers of persons of different descriptions who work or worked in the undertaking there ; or

(*b*) the disclosure by the Manpower Services Commission, the Employment Service Agency or the Training Services Agency to the Commission or the Executive of information so obtained which is of a kind specified in a notice in writing given to the disclosing body and the recipient of the information by the Secretary of State under this paragraph.

(3) In the preceding subsection any reference to a Minister of the Crown, the Commission, the Executive, the Manpower Services Commission or either of the said Agencies includes respectively a reference to an officer of his or of that body and also, in the case of a reference to the Commission, includes a reference to—

(*a*) a person performing any functions of the Commission or the Executive on its behalf by virtue of section 13(1)(*a*) ;

(*b*) an officer of a body which is so performing any such functions ; and

(*c*) an adviser appointed in pursuance of section 13(1)(*d*).

(4) A person to whom information is disclosed in pursuance of subsection (2) above shall not use the information for a purpose other than a purpose of the Commission or, as the case may be, of the Executive.

Restrictions on disclosure of information.

28.—(1) In this and the two following subsections—

(*a*) " relevant information " means information obtained by a person under section 27(1) or furnished to any person in pursuance of a requirement imposed by any of the relevant statutory provisions ; and

(*b*) " the recipient ", in relation to any relevant information, means the person by whom that information was so obtained or to whom that information was so furnished, as the case may be.

(2) Subject to the following subsection, no relevant information shall be disclosed without the consent of the person by whom it was furnished.

(3) The preceding subsection shall not apply to—

(*a*) disclosure of information to the Commission, the Executive, a government department or any enforcing authority ;

PART I

(*b*) without prejudice to paragraph (*a*) above, disclosure by the recipient of information to any person for the purpose of any function conferred on the recipient by or under any of the relevant statutory provisions;

(*c*) without prejudice to paragraph (*a*) above, disclosure by the recipient of information to—

(i) an officer of a local authority who is authorised by that authority to receive it,

(ii) an officer of a water authority or water development board who is authorised by that authority or board to receive it,

(iii) an officer of a river purification board who is authorised by that board to receive it, or

(iv) a constable authorised by a chief officer of police to receive it;

(*d*) disclosure by the recipient of information in a form calculated to prevent it from being identified as relating to a particular person or case;

(*e*) disclosure of information for the purposes of any legal proceedings or any investigation or inquiry held by virtue of section 14(2), or for the purposes of a report of any such proceedings or inquiry or of a special report made by virtue of section 14(2).

(4) In the preceding subsection any reference to the Commission, the Executive, a government department or an enforcing authority includes respectively a reference to an officer of that body or authority (including, in the case of an enforcing authority, any inspector appointed by it), and also, in the case of a reference to the Commission, includes a reference to—

(*a*) a person performing any functions of the Commission or the Executive on its behalf by virtue of section 13(1)(*a*);

(*b*) an officer of a body which is so performing any such functions; and

(*c*) an adviser appointed in pursuance of section 13(1)(*d*).

(5) A person to whom information is disclosed in pursuance of subsection (3) above shall not use the information for a purpose other than—

(*a*) in a case falling within paragraph (*a*) of that subsection, a purpose of the Commission or of the Executive or of the government department in question, or the purposes of the enforcing authority in question in connection with the relevant statutory provisions, as the case may be;

PART I

(*b*) in the case of information given to an officer of a local authority or of a water authority or of a river purification board or water development board, the purposes of the authority or board in connection with the relevant statutory provisions or any enactment whatsoever relating to public health, public safety or the protection of the environment;

(*c*) in the case of information given to a constable, the purposes of the police in connection with the relevant statutory provisions or any enactment whatsoever relating to public health, public safety or the safety of the State.

(6) In subsections (3)(*c*) and (5) above, before 16th May 1975, the references to a water authority in their application to Scotland shall be construed as references to a regional water board.

(7) A person shall not disclose any information obtained by him as a result of the exercise of any power conferred by section 14(4)(*a*) or 20 (including, in particular, any information with respect to any trade secret obtained by him in any premises entered by him by virtue of any such power) except—

(*a*) for the purposes of his functions; or

(*b*) for the purposes of any legal proceedings or any investigation or inquiry held by virtue of section 14(2) or for the purposes of a report of any such proceedings or inquiry or of a special report made by virtue of section 14(2); or

(*c*) with the relevant consent.

In this subsection " the relevant consent " means, in the case of information furnished in pursuance of a requirement imposed under section 20, the consent of the person who furnished it, and, in any other case, the consent of a person having responsibilities in relation to the premises where the information was obtained.

(8) Notwithstanding anything in the preceding subsection an inspector shall, in circumstances in which it is necessary to do so for the purpose of assisting in keeping persons (or the representatives of persons) employed at any premises adequately informed about matters affecting their health, safety and welfare, give to such persons or their representatives the following descriptions of information, that is to say—

(*a*) factual information obtained by him as mentioned in that subsection which relates to those premises or anything which was or is therein or was or is being done therein; and

PART I

(*b*) information with respect to any action which he has taken or proposes to take in or in connection with those premises in the performance of his functions ;

and, where an inspector does as aforesaid, he shall give the like information to the employer of the first-mentioned persons.

Special provisions relating to agriculture

General functions of Ministers responsible for agriculture in relation to the relevant agricultural purposes.

29.—(1) It shall be the duty of the appropriate Agriculture Minister—

(*a*) to do such things and make such arrangements as he considers appropriate for the relevant agricultural purposes ; and

(*b*) to make such arrangements as he considers appropriate for securing that employers, employees, organisations representing employers and employees respectively, and other persons concerned with matters relevant to any of those purposes are kept informed of, and adequately advised on, such matters.

(2) The Minister of Agriculture, Fisheries and Food shall make an annual report to Parliament of his proceedings under the relevant statutory provisions, and may include that report in the annual report made to Parliament in pursuance of section 13 of the Agricultural Wages Act 1948. 1948 c. 47.

(3) The Secretary of State concerned with agriculture in Scotland shall make an annual report to Parliament of his proceedings under the relevant statutory provisions.

Agricultural health and safety regulations.

30.—(1) Regulations under this section (in this Part referred to as " agricultural health and safety regulations ".) may be made for any of the relevant agricultural purposes.

(2) Agricultural health and safety regulations may be either regulations applying to Great Britain and made by the Minister of Agriculture, Fisheries and Food and the Secretary of State acting jointly, or regulations applying to England and Wales only and made by the said Minister, or regulations applying to Scotland only and made by the Secretary of State.

(3) Where health and safety regulations make provision for any purpose with respect to a matter that relates to (but not exclusively to) agricultural operations—

(*a*) provision for that purpose shall not be made with respect to that matter by agricultural health and safety regulations so as to have effect while the first-mentioned provision is in force except for the purpose of imposing requirements additional to those imposed by health and safety regulations, being additional requirements which in the opinion of the authority making the agricultural health and safety regulations are necessary or expedient in the special circumstances of agricultural operations ; and

(*b*) in the event of any inconsistency between the first-mentioned provision and any provision made with respect to that matter by agricultural health and safety regulations, the first-mentioned provision shall prevail.

(4) The provision of section 15(2) to (10) and Schedule 3 shal have effect in relation to agricultural health and safety regulations as they have effect in relation to health and safety regulations subject to the following modifications, that is to say—

(*a*) references to the relevant statutory provisions or the existing statutory provisions shall be read as references to such of those provisions as relate to agriculture ;

(*b*) in section 15(4) the references to the Commission shall be read as references to the appropriate Agriculture Minister ;

(*c*) in section 15(6) and (10) and paragraph 23 of Schedule 3, the reference to health and safety regulations shall be read as a reference to agricultural health and safety regulations.

(5) Without prejudice to the generality of subsection (1) above, agricultural health and safety regulations may, as regards agricultural licences under any of the relevant statutory provisions, make provision for requiring the authority having power to issue, renew, vary, transfer or revoke such licences to notify—

(*a*) any applicant for the issue, renewal, variation or transfer of such a licence of any proposed decision of the authority to refuse the application ; or

(*b*) the holder of such a licence of any proposed decision of the authority to revoke the licence or to vary any term, condition or restriction on or subject to which the licence is held ;

and for enabling persons aggrieved by any such proposed decision to make representations to, or to a person appointed by, the relevant authority within the period and in the manner prescribed by the regulations.

(6) In relation to any agricultural health and safety regulations made in pursuance of paragraph 2 of Schedule 3 as applied by this section, subsection (2) above shall have effect as if after the words " Great Britain " there were inserted the words " or the United Kingdom ".

PART I

Enforcement of the relevant statutory provisions in connection with agriculture.

31. Subject to any provision made by regulations under section 15, 18 or 30, it shall be the duty of the appropriate Agriculture Minister to make adequate arrangements for the enforcement of the relevant statutory provisions in their application to matters relating exclusively to the relevant agricultural purposes.

Application of provisions of this Part in connection with agriculture.

32.—(1) The following provisions of this section shall have effect with a view to the application of certain provisions of this Part in relation to the Agriculture Ministers or matters relating exclusively to the relevant agricultural purposes.

(2) Subject to the following subsection—

(*a*) sections 13, 14, 17(3), 27 and 28 shall apply in relation to the appropriate Agriculture Minister as they apply in relation to the Commission;

(*b*) section 16 shall apply in relation to matters relating exclusively to the relevant agricultural purposes as it applies in relation to other matters.

(3) In their application as provided by the preceding subsection, the provisions of this Part which are specified in the first column of Schedule 4 shall have effect subject to the modifications provided for in the second column of that Schedule.

Provisions as to offences

Offences.

33.—(1) It is an offence for a person—

(*a*) to fail to discharge a duty to which he is subject by virtue of sections 2 to 7;

(*b*) to contravene section 8 or 9;

(*c*) to contravene any health and safety regulations or agricultural health and safety regulations or any requirement or prohibition imposed under any such regulations (including any requirement or prohibition to which he is subject by virtue of the terms of or any condition or restriction attached to any licence, approval, exemption or other authority issued, given or granted under the regulations);

(*d*) to contravene any requirement imposed by or under regulations under section 14 or intentionally to obstruct any person in the exercise of his powers under that section;

(*e*) to contravene any requirement imposed by an inspector under section 20 or 25;

PART I

(*f*) to prevent or attempt to prevent any other person from appearing before an inspector or from answering any question to which an inspector may by virtue of section 20(2) require an answer ;

(*g*) to contravene any requirement or prohibition imposed by an improvement notice or a prohibition notice (including any such notice as modified on appeal) ;

(*h*) intentionally to obstruct an inspector in the exercise or performance of his powers or duties ;

(*i*) to contravene any requirement imposed by a notice under section 27(1) ;

(*j*) to use or disclose any information in contravention of section 27(4) or 28 ;

(*k*) to make a statement which he knows to be false or recklessly to make a statement which is false where the statement is made—

(i) in purported compliance with a requirement to furnish any information imposed by or under any of the relevant statutory provisions ; or

(ii) for the purpose of obtaining the issue of a document under any of the relevant statutory provisions to himself or another person ;

(*l*) intentionally to make a false entry in any register, book, notice or other document required by or under any of the relevant statutory provisions to be kept, served or given or, with intent to deceive, to make use of any such entry which he knows to be false ;

(*m*) with intent to deceive, to forge or use a document issued or authorised to be issued under any of the relevant statutory provisions or required for any purpose thereunder or to make or have in his possession a document so closely resembling any such document as to be calculated to deceive ;

(*n*) falsely to pretend to be an inspector ;

(*o*) to fail to comply with an order made by a court under section 42.

(2) A person guilty of an offence under paragraph (*d*), (*f*), (*h*) or (*n*) of subsection (1) above, or of an offence under paragraph (*e*) of that subsection consisting of contravening a requirement imposed by an inspector under section 20, shall be liable on summary conviction to a fine not exceeding £400.

(3) Subject to any provision made by virtue of section 15(6)(*d*) or by virtue of paragraph 2(2) of Schedule 3, a person guilty of an offence under any paragraph of subsection (1)

PART I

above not mentioned in the preceding subsection, or of a offence under subsection (1)(*e*) above not falling within th preceding subsection, or of an offence under any of the existin statutory provisions being an offence for which no other penalt is specified, shall be liable—

(*a*) on summary conviction, to a fine not exceeding £400

(*b*) on conviction on indictment—

(i) if the offence is one to which this sub-para graph applies, to imprisonment for a term no exceeding two years, or a fine, or both;

(ii) if the offence is not one to which the preced ing sub-paragraph applies, to a fine.

(4) Subsection (3)(*b*)(i) above applies to the followin offences—

(*a*) an offence consisting of contravening any of the relevan statutory provisions by doing otherwise than under th authority of a licence issued by the Executive or th appropriate Agriculture Minister something for th doing of which such a licence is necessary under th relevant statutory provisions;

(*b*) an offence consisting of contravening a term of or condition or restriction attached to any such licenc as is mentioned in the preceding paragraph;

(*c*) an offence consisting of acquiring or attempting t acquire, possessing or using an explosive article or sub stance (within the meaning of any of the relevan statutory provisions) in contravention of any of th relevant statutory provisions;

(*d*) an offence under subsection (1)(*g*) above consisting o contravening a requirement or prohibition impose by a prohibition notice;

(*e*) an offence under subsection (1)(*j*) above.

(5) Where a person is convicted of an offence under subsectio (1)(*g*) or (*o*) above, then, if the contravention in respect of whic he was convicted is continued after the conviction he shall (sub ject to section 42(3)) be guilty of a further offence and liable i respect thereof to a fine not exceeding £50 for each day o which the contravention is so continued.

(6) In this section " forge " has, for England and Wales, th same meaning as in the Forgery Act 1913.

1913 c. 27.

Extension of time for bringing summary proceedings.

34.—(1) Where—

(*a*) a special report on any matter to which section 14 c this Act applies is made by virtue of subsection (2)(*a* of that section; or (*b*) a report is made by the person holding an inquiry into any such matter by virtue of subsection (2)(*b*) of that section ; or PART I

(*c*) a coroner's inquest is held touching the death of any person whose death may have been caused by an accident which happened while he was at work or by a disease which he contracted or probably contracted at work or by any accident, act or omission which occurred in connection with the work of any person whatsoever ; or

(*d*) a public inquiry into any death that may have been so caused is held under the Fatal Accidents Inquiry (Scotland) Act 1895 or the Fatal Accidents and Sudden Deaths Inquiry (Scotland) Act 1906, 1895 c. 36. 1906 c. 35.

and it appears from the report or, in a case falling within paragraph (*c*) or (*d*) above, from the proceedings at the inquest or inquiry, that any of the relevant statutory provisions was contravened at a time which is material in relation to the subject-matter of the report, inquest or inquiry, summary proceedings against any person liable to be proceeded against in respect of the contravention may be commenced at any time within three months of the making of the report or, in a case falling within paragraph (*c*) or (*d*) above, within three months of the conclusion of the inquest or inquiry.

(2) Where an offence under any of the relevant statutory provisions is committed by reason of a failure to do something at or within a time fixed by or under any of those provisions, the offence shall be deemed to continue until that thing is done.

(3) Summary proceedings for an offence to which this subsection applies may be commenced at any time within six months from the date on which there comes to the knowledge of a responsible enforcing authority evidence sufficient in the opinion of that authority to justify a prosecution for that offence ; and for the purposes of this subsection—

(*a*) a certificate of an enforcing authority stating that such evidence came to its knowledge on a specified date shall be conclusive evidence of that fact ; and

(*b*) a document purporting to be such a certificate and to be signed by or on behalf of the enforcing authority in question shall be presumed to be such a certificate unless the contrary is proved.

(4) The preceding subsection applies to any offence under any of the relevant statutory provisions which a person commits by virtue of any provision or requirement to which he is subject as the designer, manufacturer, importer or supplier of anything ;

PART I

and in that subsection " responsible enforcing authority " means an enforcing authority within whose field of responsibility the offence in question lies, whether by virtue of section 35 or otherwise.

(5) In the application of subsection (3) above to Scotland—

(*a*) for the words from " there comes " to " that offence " there shall be substituted the words " evidence, sufficient in the opinion of the enforcing authority to justify a report to the Lord Advocate with a view to consideration of the question of prosecution, comes to the knowledge of the authority " ;

(*b*) at the end of paragraph (*b*) there shall be added the words " and

1954 c. 48.

(*c*) section 23(2) of the Summary Jurisdiction (Scotland) Act 1954 (date of commencement of proceedings) shall have effect as it has effect for the purposes of that section.".

Venue.

35. An offence under any of the relevant statutory provisions committed in connection with any plant or substance may, if necessary for the purpose of bringing the offence within the field of responsibility of any enforcing authority or conferring jurisdiction on any court to entertain proceedings for the offence, be treated as having been committed at the place where that plant or substance is for the time being.

Offences due to fault of other person.

36.—(1) Where the commission by any person of an offence under any of the relevant statutory provisions is due to the act or default of some other person, that other person shall be guilty of the offence, and a person may be charged with and convicted of the offence by virtue of this subsection whether or not proceedings are taken against the first-mentioned person.

(2) Where there would be or have been the commission of an offence under section 33 by the Crown but for the circumstance that that section does not bind the Crown, and that fact is due to the act or default of a person other than the Crown that person shall be guilty of the offence which, but for that circumstance, the Crown would be committing or would have committed, and may be charged with and convicted of that offence accordingly.

(3) The preceding provisions of this section are subject to any provision made by virtue of section 15(6).

Offences by bodies corporate.

37.—(1) Where an offence under any of the relevant statutory provisions committed by a body corporate is proved to have been committed with the consent or connivance of, or to have been attributable to any neglect on the part of, any director, manager, secretary or other similar officer of the body corporate or a person who was purporting to act in any such capacity, he as well as the body corporate shall be guilty of that offence and shall be liable to be proceeded against and punished accordingly.

(2) Where the affairs of a body corporate are managed by its members, the preceding subsection shall apply in relation to the acts and defaults of a member in connection with his functions of management as if he were a director of the body corporate.

Restriction on institution of proceedings in England and Wales.

38. Proceedings for an offence under any of the relevant statutory provisions shall not, in England and Wales, be instituted except by an inspector or by or with the consent of the Director of Public Prosecutions.

Prosecutions by inspectors.

39.—(1) An inspector, if authorised in that behalf by the enforcing authority which appointed him, may, although not of counsel or a solicitor, prosecute before a magistrates' court proceedings for an offence under any of the relevant statutory provisions.

(2) This section shall not apply to Scotland.

Onus of proving limits of what is practicable etc.

40. In any proceedings for an offence under any of the relevant statutory provisions consisting of a failure to comply with a duty or requirement to do something so far as is practicable or so far as is reasonably practicable, or to use the best practicable means to do something, it shall be for the accused to prove (as the case may be) that it was not practicable or not reasonably practicable to do more than was in fact done to satisfy the duty or requirement, or that there was no better practicable means than was in fact used to satisfy the duty or requirement.

Evidence.

41.—(1) Where an entry is required by any of the relevant statutory provisions to be made in any register or other record, the entry, if made, shall, as against the person by or on whose behalf it was made, be admissible as evidence or in Scotland sufficient evidence of the facts stated therein.

(2) Where an entry which is so required to be so made with respect to the observance of any of the relevant statutory provisions has not been made, that fact shall be admissible as evidence or in Scotland sufficient evidence that that provision has not been observed.

PART I

Power of court to order cause of offence to be remedied or, in certain cases, forfeiture.

42.—(1) Where a person is convicted of an offence under any of the relevant statutory provisions in respect of any matters which appear to the court to be matters which it is in his power to remedy, the court may, in addition to or instead of imposing any punishment, order him, within such time as may be fixed by the order, to take such steps as may be specified in the order for remedying the said matters.

(2) The time fixed by an order under subsection (1) above may be extended or further extended by order of the court on an application made before the end of that time as originally fixed or as extended under this subsection, as the case may be.

(3) Where a person is ordered under subsection (1) above to remedy any matters, that person shall not be liable under any of the relevant statutory provisions in respect of those matters in so far as they continue during the time fixed by the order or any further time allowed under subsection (2) above.

(4) Subject to the following subsection, the court by or before which a person is convicted of an offence such as is mentioned in section 33(4)(*c*) in respect of any such explosive article or substance as is there mentioned may order the article or substance in question to be forfeited and either destroyed or dealt with in such other manner as the court may order.

(5) The court shall not order anything to be forfeited under the preceding subsection where a person claiming to be the owner of or otherwise interested in it applies to be heard by the court, unless an opportunity has been given to him to show cause why the order should not be made.

Financial provisions

Financial provisions.

43.—(1) It shall be the duty of the Secretary of State to pay to the Commission such sums as are approved by the Treasury and as he considers appropriate for the purpose of enabling the Commission to perform its functions; and it shall be the duty of the Commission to pay to the Executive such sums as the Commission considers appropriate for the purpose of enabling the Executive to perform its functions.

(2) Regulations may provide for such fees as may be fixed by or determined under the regulations to be payable for or in connection with the performance by or on behalf of any authority to which this subsection applies of any function conferred on that authority by or under any of the relevant statutory provisions.

(3) Subsection (2) above applies to the following authorities, namely the Commission, the Executive, the Secretary of State,

the Minister of Agriculture, Fisheries and Food, every enforcing authority, and any other person on whom any function is conferred by or under any of the relevant statutory provisions. PART I

(4) Regulations under this section may specify the person by whom any fee payable under the regulations is to be paid; but no such fee shall be made payable by a person in any of the following capacieies, namely an employee, a person seeking employment, a person training for employment, and a person seeking training for employment.

(5) Without prejudice to section 82(3), regulations under this section may fix or provide for the determination of different fees in relation to different functions, or in relation to the same function in different circumstances.

(6) The power to make regulations under this section shall be exercisable—

(*a*) as regards functions with respect to matters not relating exclusively to agricultural operations, by the Secretary of State;

(*b*) as regards functions with respect to matters relating exclusively to the relevant agricultural purposes, by the appropriate agricultural authority.

(7) Regulations under this section as regards functions falling within subsection (6)(*b*) above may be either regulations applying to Great Britain and made by the Minister of Agriculture, Fisheries and Food and the Secretary of State acting jointly, or regulations applying to England and Wales only and made by the said Minister, or regulations applying to Scotland only and made by the Secretary of State; and in subsection (6)(*b*) above " the appropriate agricultural authority " shall be construed accordingly.

(8) In subsection (4) above the references to a person training for employment and a person seeking training for employment shall include respectively a person attending an industrial rehabilitation course provided by virtue of the Employment and Training Act 1973 and a person seeking to attend such a course. 1973 c. 50.

(9) For the purposes of this section the performance by an inspector of his functions shall be treated as the performance by the enforcing authority which appointed him of functions conferred on that authority by or under any of the relevant statutory provisions.

PART I

Miscellaneous and supplementary

Appeals in connection with licensing provisions in the relevant statutory provisions.

44.—(1) Any person who is aggrieved by a decision of an authority having power to issue licences (other than agricultural licences and nuclear site licences) under any of the relevant statutory provisions—

(*a*) refusing to issue him a licence, to renew a licence held by him, or to transfer to him a licence held by another;

(*b*) issuing him a licence on or subject to any term, condition or restriction whereby he is aggrieved;

(*c*) varying or refusing to vary any term, condition or restriction on or subject to which a licence is held by him; or

(*d*) revoking a licence held by him,

may appeal to the Secretary of State.

(2) The Secretary of State may, in such cases as he considers it appropriate to do so, having regard to the nature of the questions which appear to him to arise, direct that an appeal under this section shall be determined on his behalf by a person appointed by him for that purpose.

(3) Before the determination of an appeal the Secretary of State shall ask the appellant and the authority against whose decision the appeal is brought whether they wish to appear and be heard on the appeal and—

(*a*) the appeal may be determined without a hearing of the parties if both of them express a wish not to appear and be heard as aforesaid;

(*b*) the Secretary of State shall, if either of the parties expresses a wish to appear and be heard, afford to both of them an opportunity of so doing.

1971 c. 62.

(4) The Tribunals and Inquiries Act 1971 shall apply to a hearing held by a person appointed in pursuance of subsection (2) above to determine an appeal as it applies to a statutory inquiry held by the Secretary of State, but as if in section 12(1) of that Act (statement of reasons for decisions) the reference to any decision taken by the Secretary of State included a reference to a decision taken on his behalf by that person.

(5) A person who determines an appeal under this section on behalf of the Secretary of State and the Secretary of State, if he determines such an appeal, may give such directions as he considers appropriate to give effect to his determination.

(6) The Secretary of State may pay to any person appointed to hear or determine an appeal under this section on his behalf such remuneration and allowances as the Secretary of State may with the approval of the Minister for the Civil Service determine.

PART I

(7) In this section—

(*a*) " licence " means a licence under any of the relevant statutory provisions other than an agricultural licence or nuclear site licence ;

(*b*) " nuclear site licence " means a licence to use a site for the purpose of installing or operating a nuclear installation within the meaning of the following subsection.

(8) For the purposes of the preceding subsection " nuclear installation " means—

(*a*) a nuclear reactor (other than such a reactor comprised in a means of transport, whether by land, water or air) ; or

(*b*) any other installation of such class or description as may be prescribed for the purposes of this paragraph or section 1(1)(*b*) of the Nuclear Installations Act 1965, being an installation designed or adapted for— 1965 c. 57.

(i) the production or use of atomic energy ; or

(ii) the carrying out of any process which is preparatory or ancillary to the production or use of atomic energy and which involves or is capable of causing the emission of ionising radiations ; or

(iii) the storage, processing or disposal of nuclear fuel or of bulk quantities of other radioactive matter, being matter which has been produced or irradiated in the course of the production or use of nuclear fuel ;

and in this subsection—

" atomic energy " has the meaning assigned by the Atomic Energy Act 1946 ; 1946 c. 80.

" nuclear reactor " means any plant (including any machinery, equipment or appliance, whether affixed to land or not) designed or adapted for the production of atomic energy by a fission process in which a controlled chain reaction can be maintained without an additional source of neutrons.

Default powers.

45.—(1) Where, in the case of a local authority who are an enforcing authority, the Commission is of the opinion that an investigation should be made as to whether that local authority have failed to perform any of their enforcement functions the Commission may make a report to the Secretary of State.

(2) The Secretary of State may, after considering a report submitted to him under the preceding subsection, cause a local inquiry to be held ; and the provisions of subsections (2) to (5) of section 250 of the Local Government Act 1972 as to local inquiries shall, without prejudice to the generality of subsection (1) of that section, apply to a local inquiry so held 1972 c. 70.

PART I as they apply to a local inquiry held in pursuance of that section.

(3) If the Secretary of State is satisfied, after having caused a local inquiry to be held into the matter, that a local authority have failed to perform any of their enforcement functions, he may make an order declaring the authority to be in default.

(4) An order made by virtue of the preceding subsection which declares an authority to be in default may, for the purpose of remedying the default, direct the authority (hereafter in this section referred to as " the defaulting authority ") to perform such of their enforcement functions as are specified in the order in such manner as may be so specified and may specify the time or times within which those functions are to be performed by the authority.

(5) If the defaulting authority fail to comply with any direction contained in such an order the Secretary of State may, instead of enforcing the order by mandamus, make an order transferring to the Executive such of the enforcement functions of the defaulting authority as he thinks fit.

(6) Where any enforcement functions of the defaulting authority are transferred in pursuance of the preceding subsection, the amount of any expenses which the Executive certifies were incurred by it in performing those functions shall on demand be paid to it by the defaulting authority.

(7) Any expenses which in pursuance of the preceding subsection are required to be paid by the defaulting authority in respect of any enforcement functions transferred in pursuance of this section shall be defrayed by the authority in the like manner, and shall be debited to the like account, as if the enforcement functions had not been transferred and the expenses had been incurred by the authority in performing them.

(8) Where the defaulting authority are required to defray any such expenses the authority shall have the like powers for the purpose of raising the money for defraying those expenses as they would have had for the purpose of raising money required for defraying expenses incurred for the purpose of the enforcement functions in question.

(9) An order transferring any enforcement functions of the defaulting authority in pursuance of subsection (5) above may provide for the transfer to the Executive of such of the rights, liabilities and obligations of the authority as the Secretary of State considers appropriate ; and where such an order is revoked the Secretary of State may, by the revoking order or a

subsequent order, make such provision as he considers appropriate with respect to any rights, liabilities and obligations held by the Executive for the purposes of the transferred enforcement functions. PART I

(10) The Secretary of State may by order vary or revoke any order previously made by him in pursuance of this section.

(11) In this section " enforcement functions ", in relation to a local authority, means the functions of the authority as an enforcing authority.

(12) In the application of this section to Scotland—

(*a*) in subsection (2) for the words " subsections (2) to (5) of section 250 of the Local Government Act 1972 " there shall be substituted the words " subsections (2) to (8) of section 210 of the Local Government (Scotland) Act 1973 ", except that before 16th May 1975 for the said words there shall be substituted the words " subsections (2) to (9) of section 355 of the Local Government (Scotland) Act 1947 "; 1972 c. 70. 1973 c. 65. 1947 c. 43.

(*b*) in subsection (5) the words " instead of enforcing the order by mandamus " shall be omitted.

Service of notices.

46.—(1) Any notice required or authorised by any of the relevant statutory provisions to be served on or given to an inspector may be served or given by delivering it to him or by leaving it at, or sending it by post to, his office.

(2) Any such notice required or authorised to be served on or given to a person other than an inspector may be served or given by delivering it to him, or by leaving it at his proper address, or by sending it by post to him at that address.

(3) Any such notice may—

(*a*) in the case of a body corporate, be served on or given to the secretary or clerk of that body;

(*b*) in the case of a partnership, be served on or given to a partner or a person having the control or management of the partnership business or, in Scotland, the firm.

(4) For the purposes of this section and of section 26 of the Interpretation Act 1889 (service of documents by post) in its application to this section, the proper address of any person on or to whom any such notice is to be served or given shall be his last known address, except that— 1889 c. 63.

(*a*) in the case of a body corporate or their secretary or clerk, it shall be the address of the registered or principal office of that body;

PART I

(*b*) in the case of a partnership or a person having the control or the management of the partnership business, it shall be the principal office of the partnership;

and for the purposes of this subsection the principal office of a company registered outside the United Kingdom or of a partnership carrying on business outside the United Kingdom shall be their principal office within the United Kingdom.

(5) If the person to be served with or given any such notice has specified an address within the United Kingdom other than his proper address within the meaning of subsection (4) above as the one at which he or someone on his behalf will accept notices of the same description as that notice, that address shall also be treated for the purposes of this section and section 26 of the Interpretation Act 1889 as his proper address.

1889 c. 63.

(6) Without prejudice to any other provision of this section, any such notice required or authorised to be served on or given to the owner or occupier of any premises (whether a body corporate or not) may be served or given by sending it by post to him at those premises, or by addressing it by name to the person on or to whom it is to be served or given and delivering it to some responsible person who is or appears to be resident or employed in the premises.

(7) If the name or the address of any owner or occupier of premises on or to whom any such notice as aforesaid is to be served or given cannot after reasonable inquiry be ascertained, the notice may be served or given by addressing it to the person on or to whom it is to be served or given by the description of "owner" or "occupier" of the premises (describing them) to which the notice relates, and by delivering it to some responsible person who is or appears to be resident or employed in the premises, or, if there is no such person to whom it can be delivered, by affixing it or a copy of it to some conspicuous part of the premises.

(8) The preceding provisions of this section shall apply to the sending or giving of a document as they apply to the giving of a notice.

Civil liability.

47.—(1) Nothing in this Part shall be construed—

(*a*) as conferring a right of action in any civil proceedings in respect of any failure to comply with any duty imposed by sections 2 to 7 or any contravention of section 8; or

(*b*) as affecting the extent (if any) to which breach of a duty imposed by any of the existing statutory provisions is actionable; or

PART I

(c) as affecting the operation of section 12 of the Nuclear Installations Act 1965 (right to compensation by virtue of certain provisions of that Act). 1965 c. 57.

(2) Breach of a duty imposed by health and safety regulations or agricultural health and safety regulations shall, so far as it causes damage, be actionable except in so far as the regulations provide otherwise.

(3) No provision made by virtue of section 15(6)(*b*) shall afford a defence in any civil proceedings, whether brought by virtue of subsection (2) above or not; but as regards any duty imposed as mentioned in subsection (2) above health and safety regulations or, as the case may be, agricultural health and safety regulations may provide for any defence specified in the regulations to be available in any action for breach of that duty.

(4) Subsections (1)(*a*) and (2) above are without prejudice to any right of action which exists apart from the provisions of this Act, and subsection (3) above is without prejudice to any defence which may be available apart from the provisions of the regulations there mentioned.

(5) Any term of an agreement which purports to exclude or restrict the operation of subsection (2) above, or any liability arising by virtue of that subsection shall be void, except in so far as health and safety regulations or, as the case may be, agricultural health and safety regulations provide otherwise.

(6) In this section "damage" includes the death of, or injury to, any person (including any disease and any impairment of a person's physical or mental condition).

Application to Crown.

48.—(1) Subject to the provisions of this section, the provisions of this Part, except sections 21 to 25 and 33 to 42, and of regulations made under this Part shall bind the Crown.

(2) Although they do not bind the Crown, sections 33 to 42 shall apply to persons in the public service of the Crown as they apply to other persons.

(3) For the purposes of this Part and regulations made thereunder persons in the service of the Crown shall be treated as employees of the Crown whether or not they would be so treated apart from this subsection.

(4) Without prejudice to section 15(5), the Secretary of State may, to the extent that it appears to him requisite or expedient to do so in the interests of the safety of the State or the safe custody of persons lawfully detained, by order exempt the Crown either generally or in particular respects from all or any of the provisions of this Part which would, by virtue of subsection (1) above, bind the Crown.

PART I

(5) The power to make orders under this section shall be exercisable by statutory instrument, and any such order may be varied or revoked by a subsequent order.

1947 c. 44.

(6) Nothing in this section shall authorise proceedings to be brought against Her Majesty in her private capacity, and this subsection shall be construed as if section 38(3) of the Crown Proceedings Act 1947 (interpretation of references in that Act to Her Majesty in her private capacity) were cantained in this Act.

Adaptation of enactments to metric units or appropriate metric units.

49.—(1) The appropriate Minister may by regulations amend—

(*a*) any of the relevant statutory provisions ; or

(*b*) any provision of an enactment which relates to any matter relevant to any of the general purposes of this Part but is not among the relevant statutory provisions ; or

(*c*) any provision of an instrument made or having effect under any such enactment as is mentioned in the preceding paragraph,

by substituting an amount or quantity expressed in metric units for an amount or quantity not so expressed or by substituting an amount or quanity expressed in metric units of a description specified in the regulations for an amount or quantity expressed in metric units of a different description.

(2) The amendments shall be such as to preserve the effect of the provisions mentioned except to such extent as in the opinion of the appropriate Minister is necessary to obtain amounts expressed in convenient and suitable terms.

(3) Regulations made by the appropriate Minister under this subsection may, in the case of a provision which falls within any of paragraphs (*a*) to (*c*) of subsection (1) above and contains words which refer to units other than metric units, repeal those words if the appropriate Minister is of the opinion that those words could be omitted without altering the effect of that provision.

(4) In this section the appropriate Minister means—

(*a*) in relation to any provision not relating exclusively to agricultural operations the Secretary of State ;

(*b*) in relation to any provision relating exclusively to the relevant agricultural purposes that applies to Great Britain or the United Kingdom the Agriculture Ministers ;

(*c*) in relation to any provision so relating that applies to England and Wales only, the Minister of Agriculture, Fisheries and Food ;

(*d*) in relation to any provision so relating that applies to Scotland only, the Secretary of State.

Regulations under the relevant statutory provisions.

50.—(1) Subject to subsection (5) below any power to make regulations conferred on the Secretary of State by any of the relevant statutory provisions may be exercised by him either so as to give effect (with or without modifications) to proposals for the making of regulations by him under that power submitted to him by the Commission or independently of any such proposals, but before making any regulations under any of those provisions independently of any such proposals the Secretary of State shall consult the Commission and such other bodies as appear to him to be appropriate.

(2) Where the Secretary of State proposes to exercise any such power as is mentioned in the preceding subsection so as to give effect to any such proposals as are there mentioned with modifications, he shall, before making the regulations, consult the Commission.

(3) Where the Commission proposes to submit to the Secretary of State any such proposals as are mentioned in subsection (1) above except proposals for the making of regulations under section 43(2), it shall, before so submitting them, consult—

(*a*) any government department or other body that appears to the Commission to be appropriate (and, in particular, in the case of proposals for the making of regulations under section 18(2), any body representing local authorities that so appears, and, in the case of proposals for the making of regulations relating to electro-magnetic radiations, the National Radiological Protection Board) ;

(*b*) such government departments and other bodies, if any, as, in relation to any matter dealt with in the proposals, the Commission is required to consult under this subsection by virtue of directions given to it by the Secretary of State.

(4) Where the Minister of Agriculture, Fisheries and Food and the Secretary of State or either of them propose or proposes to make any regulations under any of the relevant statutory provisions, they or he shall before making the regulations consult the Commission and such other bodies as appear to them or him to be appropriate.

(5) Subsections (1) to (3) above shall not apply to any power of the Secretary of State to make regulations which is capable

PART I

of being exercised by him for Great Britain jointly with the Minister of Agriculture, Fisheries and Food.

Exclusion of application to domestic employment.

51. Nothing in this Part shall apply in relation to a person by reason only that he employs another, or is himself employed, as a domestic servant in a private household.

Meaning of work and at work.

52.—(1) For the purposes of this Part—

(*a*) " work " means work as an employee or as a self-employed person;

(*b*) an employee is at work throughout the time when he is in the course of his employment, but not otherwise; and

(*c*) a self-employed person is at work throughout such time as he devotes to work as a self-employed person;

and, subject to the following subsection, the expressions " work " and " at work ", in whatever context, shall be construed accordingly.

(2) Regulations made under this subsection may—

(*a*) extend the meaning of " work " and " at work " for the purposes of this Part; and

(*b*) in that connection provide for any of the relevant statutory provisions to have effect subject to such adaptations as may be specified in the regulations.

(3) The power to make regulations under subsection (2) above shall be exercisable—

(*a*) in relation to activities not relating exclusively to agricultural operations, by the Secretary of State;

(*b*) in relation to activities relating exclusively to the relevant agricultural purposes, by the appropriate agriculture authority.

(4) Regulations under subsection (2) above in relation to activities falling within subsection (3)(*b*) above may be either regulations applying to Great Britain and made by the Minister of Agriculture, Fisheries and Food and the Secretary of State acting jointly, or regulations applying to England and Wales only and made by the said Minister, or regulations applying to Scotland only and made by the Secretary of State; and in subsection (3)(*b*) above " the appropriate agriculture authority " shall be construed accordingly.

General interpretation of Part I.

53.—(1) In this Part, unless the context otherwise requires—

" agriculture ", subject to subsection (3) below, includes horticulture, fruit growing, seed growing, dairy farming, livestock breeding and keeping (including the management of livestock up to the point of slaughter or export from Great Britain), forestry, the use of land as grazing land, meadow land, osier land, market gardens and nursery grounds, and the preparation of land for agricultural use, and " agricultural " shall be construed accordingly ;

" the Agriculture Ministers " means the Minister of Agriculture, Fisheries and Food and the Secretary of State and, in the case of anything falling to be done by the Agriculture Ministers, means those Ministers acting jointly ;

" agricultural health and safety regulations " has the meaning assigned by section 30(1) ;

" agricultural licence " means a licence of the Agriculture Ministers or either of them under any of the relevant statutory provisions ;

" agricultural operation " does not include an agricultural operation performed otherwise than in the course of a trade, business or other undertaking (whether carried on for profit or not) but, subject to subsection (2) below, includes any operation incidental to agriculture which is performed in the course of such a trade, business or undertaking ;

" the appropriate Agriculture Minister " means, for the purpose of the application of any of the relevant statutory provisions to England and Wales, the Minister of Agriculture, Fisheries and Food, and, for the purpose of the application of any of those provisions to Scotland, the Secretary of State ;

" article for use at work " means—

(*a*) any plant designed for use or operation (whether exclusively or not) by persons at work, and

(*b*) any article designed for use as a component in any such plant ;

" code of practice " (without prejudice to section 16(8)) includes a standard, a specification and any other documentary form of practical guidance ;

" the Commission " has the meaning assigned by section 10(2) ;

" conditional sale agreement " means an agreement for the sale of goods under which the purchase price or part of it is payable by instalments, and the property in the goods is to remain in the seller (notwithstanding that the buyer is to be in possession of the goods)

PART I

until such conditions as to the payment of instalments or otherwise as may be specified in the agreement are fulfilled ;

" contract of employment " means a contract of employment or apprenticeship (whether express or implied and, if express, whether oral or in writing) ;

" credit-sale agreement " means an agreement for the sale of goods, under which the purchase price or part of it is payable by instalments, but which is not a conditional sale agreement ;

" domestic premises " means premises occupied as a private dwelling (including any garden, yard, garage, outhouse or other appurtenance of such premises which is not used in common by the occupants of more than one such dwelling), and " non-domestic premises " shall be construed accordingly ;

" employee " means an individual who works under a contract of employment, and related expressions shall be construed accordingly ;

" enforcing authority " has the meaning assigned by section 18(7) ;

" the Executive " has the meaning assigned by section 10(5) ;

" the existing statutory provisions " means the following provisions while and to the extent that they remain in force, namely the provisions of the Acts mentioned in Schedule 1 which are specified in the third column of that Schedule and of the regulations, orders or other instruments of a legislative character made or having effect under any provision so specified ;

" forestry " includes—

(*a*) the felling of trees and the extraction and primary conversion of trees within the wood or forest in which they were grown, and

(*b*) the use of land for woodlands where that use is ancillary to the use of land for other agricultural purposes ;

" the general purposes of this Part " has the meaning assigned by section 1 ;

" health and safety regulations " has the meaning assigned by section 15(1) ;

" hire-purchase agreement " means an agreement other than a conditional sale agreement, under which—

(*a*) goods are bailed or (in Scotland) hired in return for periodical payments by the person to whom they are bailed or hired ; and (*b*) the property in the goods will pass to that person if the terms of the agreement are complied with and one or more of the following occurs :

(i) the exercise of an option to purchase by that person ;

(ii) the doing of any other specified act by any party to the agreement ;

(iii) the happening of any other event ;

and " hire-purchase " shall be construed accordingly ;

" improvement notice " means a notice under section 21 ;

" inspector " means an inspector appointed under section 19 ;

" livestock " includes any creature kept for the production of food, wool, skins or fur, or for the purpose of its use in the carrying on of any agricultural activity ;

" local authority " means—

(*a*) in relation to England and Wales, a county council, the Greater London Council, a district council, a London borough council, the Common Council of the City of London, the Sub-Treasurer of the Inner Temple, or the Under-Treasurer of the Middle Temple,

(*b*) in relation to Scotland, a regional, islands or district council except that before 16th May 1975 it means a town council or county council ;

" offshore installation " means any installation which is intended for underwater exploitation of mineral resources or exploration with a view to such exploitation ;

" personal injury " includes any disease and any impairment of a person's physical or mental condition ;

" plant " includes any machinery, equipment or appliance ;

" premises " includes any place and, in particular, includes—

(*a*) any vehicle, vessel, aircraft or hovercraft,

(*b*) any installation on land (including the foreshore and other land intermittently covered by water), any offshore installation, and any other installation (whether floating, or resting on the seabed or the subsoil thereof, or resting on other land covered with water or the subsoil thereof), and

(*c*) any tent or movable structure ;

" prescribed " means prescribed by regulations made by the Secretary of State ;

PART I

"prohibition notice" means a notice under section 22;

"the relevant agricultural purposes" means the following purposes, that is to say—

(*a*) securing the health, safety and welfare at work of persons engaged in agricultural operations,

(*b*) protecting persons other than persons so engaged against risks to health or safety arising out of or in connection with the activities at work of persons so engaged;

and the reference in paragraph (*b*) above to the risks there mentioned shall be construed in accordance with section 1(3);

"the relevant statutory provisions" means—

(*a*) the provisions of this Part and of any health and safety regulations and agricultural health and safety regulations; and

(*b*) the existing statutory provisions;

"self-employed person" means an individual who works for gain or reward otherwise than under a contract of employment, whether or not he himself employs others;

"substance" means any natural or artificial substance, whether in solid or liquid form or in the form of a gas or vapour;

"substance for use at work" means any substance intended for use (whether exclusively or not) by persons at work;

"supply", where the reference is to supplying articles or substances, means supplying them by way of sale, lease, hire or hire-purchase, whether as principal or agent for another.

(2) In determining in any particular case whether an operation is incidental to agriculture within the meaning of the definition of "agricultural operation" in the preceding subsection, regard shall be had to the magnitude of the operation and to the scale on which it is performed as well as to all other relevant circumstances.

(3) Provision may be made by order for directing that for the purposes of this Part any activity or operation specified in the order which would or would not otherwise be agriculture within the meaning of this Part shall be treated as not being or, as the case may be, being agriculture for those purposes.

(4) An order under subsection (3) above may be either an order applying to Great Britain and made by the Minister of Agriculture, Fisheries and Food and the Secretary of State acting

jointly, or an order applying to England and Wales only and made by the said Minister, or an order applying to Scotland only and made by the Secretary of State.

(5) An order under subsection (3) above may be varied or revoked by a subsequent order thereunder made by the authority who made the original order.

(6) The power to make orders under subsection (3) above shall be exercisable by statutory instrument subject to annulment in pursuance of a resolution of either House of Parliament.

Application of Part I to Isles of Scilly.

54. This Part, in its application to the Isles of Scilly, shall apply as if those Isles were a local government area and the Council of those Isles were a local authority.

Part II

The Employment Medical Advisory Service

Functions of, and responsibility for maintaining, employment medical advisory service.

55.—(1) There shall continue to be an employment medical advisory service, which shall be maintained for the following purposes, that is to say—

(*a*) securing that the Secretary of State, the Health and Safety Commission, the Manpower Services Commission and others concerned with the health of employed persons or of persons seeking or training for employment can be kept informed of, and adequately advised on, matters of which they ought respectively to take cognisance concerning the safeguarding and improvement of the health of those persons;

(*b*) giving to employed persons and persons seeking or training for employment information and advice on health in relation to employment and training for employment;

(*c*) other purposes of the Secretary of State's functions relating to employment.

(2) The authority responsible for maintaining the said service shall be the Secretary of State; but if arrangements are made by the Secretary of State for that responsibility to be discharged on his behalf by the Health and Safety Commission or some other body, then, while those arrangements operate, the body so discharging that responsibility (and not the Secretary of State) shall be the authority responsible for maintaining that service.

PART II

(3) The authority for the time being responsible for maintaining the said service may also for the purposes mentioned in subsection (1) above, and for the purpose of assisting employment medical advisers in the performance of their functions, investigate or assist in, arrange for or make payments in respect of the investigation of problems arising in connection with any such matters as are so mentioned or otherwise in connection with the functions of employment medical advisers, and for the purpose of investigating or assisting in the investigation of such problems may provide and maintain such laboratories and other services as appear to the authority to be requisite.

(4) Any arrangements made by the Secretary of State in pursuance of subsection (2) above may be terminated by him at any time, but without prejudice to the making of other arrangements at any time in pursuance of that subsection (including arrangements which are to operate from the time when any previous arrangements so made cease to operate).

(5) Without prejudice to sections 11(4)(*a*) and 12(*b*), it shall be the duty of the Health and Safety Commission, if so directed by the Secretary of State, to enter into arrangements with him for the Commission to be responsible for maintaining the said service.

(6) In subsection (1) above—

(*a*) the reference to persons training for employment shall include persons attending industrial rehabilitation courses provided by virtue of the Employment and Training Act 1973; and

1973 c. 50.

(*b*) the reference to persons (other than the Secretary of State and the Commissions there mentioned) concerned with the health of employed persons or of persons seeking or training for employment shall be taken to include organisations representing employers, employees and occupational health practitioners respectively.

Functions of authority responsible for maintaining the service.

56.—(1) The authority for the time being responsible for maintaining the employment medical advisory service shall for the purpose of discharging that responsibility appoint persons to be employment medical advisers, and may for that purpose appoint such other officers and servants as it may determine, subject however to the requisite approval as to numbers, that is to say—

(*a*) where that authority is the Secretary of State, the approval of the Minister for the Civil Service;

(*b*) otherwise, the approval of the Secretary of State given with the consent of that Minister.

PART II

(2) A person shall not be qualified to be appointed, or to be, an employment medical adviser unless he is a fully registered medical practitioner.

(3) The authority for the time being responsible for maintaining the said service may determine the cases and circumstances in which the employment medical advisers or any of them are to perform the duties or exercise the powers conferred on employment medical advisers by or under this Act or otherwise.

(4) Where as a result of arrangements made in pursuance of section 55(2) the authority responsible for maintaining the said service changes, the change shall not invalidate any appointment previously made under subsection (1) above, and any such appointment subsisting when the change occurs shall thereafter have effect as if made by the new authority.

Fees.

57.—(1) The Secretary of State may by regulations provide for such fees as may be fixed by or determined under the regulations to be payable for or in connection with the performance by the authority responsible for maintaining the employment medical advisory service of any function conferred for the purposes of that service on that authority by virtue of this Part or otherwise.

(2) For the purposes of this section, the performance by an employment medical adviser of his functions shall be treated as the performance by the authority responsible for maintaining the said service of functions conferred on that authority as mentioned in the preceding subsection.

(3) The provisions of subsections (4), (5) and (8) of section 43 shall apply in relation to regulations under this section with the modification that references to subsection (2) of that section shall be read as references to subsection (1) of this section.

(4) Where an authority other than the Secretary of State is responsible for maintaining the said service, the Secretary of State shall consult that authority before making any regulations under this section.

Other financial provisions.

58.—(1) The authority for the time being responsible for maintaining the employment medical advisory service may pay—

(*a*) to employment medical advisers such salaries or such fees and travelling or other allowances; and

(*b*) to other persons called upon to give advice in connection with the execution of the authority's functions under

PART II

this Part such travelling or other allowances or compensation for loss of remunerative time; and

(*c*) to persons attending for medical examinations conducted by, or in accordance with arrangements made by, employment medical advisers (including pathological, physiological and radiological tests and similar investigations so conducted) such travelling or subsistence allowances or such compensation for loss of earnings,

as the authority may, with the requisite approval, determine.

(2) For the purposes of the preceding subsection the requisite approval is—

(*a*) where the said authority is the Secretary of State, the approval of the Minister for the Civil Service;

(*b*) otherwise, the approval of the Secretary of State given with the consent of that Minister.

(3) Where an authority other than the Secretary of State is responsible for maintaining the said service, it shall be the duty of the Secretary of State to pay to that authority such sums as are approved by the Treasury and as he considers appropriate for the purpose of enabling the authority to discharge that responsibility.

Duty of responsible authority to keep accounts and to report.

59.—(1) It shall be the duty of the authority for the time being responsible for maintaining the employment medical advisory service—

(*a*) to keep, in relation to the maintenance of that service, proper accounts and proper records in relation to the accounts;

(*b*) to prepare in respect of each accounting year a statement of accounts relating to the maintenance of that service in such form as the Secretary of State may direct with the approval of the Treasury; and

(*c*) to send copies of the statement to the Secretary of State and the Comptroller and Auditor General before the end of the month of November next following the accounting year to which the statement relates.

(2) The Comptroller and Auditor General shall examine, certify and report on each statement received by him in pursuance of subsection (1) above and shall lay copies of each statement and of his report before each House of Parliament.

(3) It shall also be the duty of the authority responsible for maintaining the employment medical advisory service to make to the Secretary of State, as soon as possible after the end of

each accounting year, a report on the discharge of its responsibilities in relation to that service during that year; and the Secretary of State shall lay before each House of Parliament a copy of each report made to him in pursuance of this subsection. PART II

(4) Where as a result of arrangements made in pursuance of section 55(2) the authority responsible for maintaining the employment medical advisory service changes, the change shall not affect any duty imposed by this section on the body which was responsible for maintaining that service before the change.

(5) No duty imposed on the authority for the time being responsible for maintaining the employment medical advisory service by subsection (1) or (3) above shall fall on the Commission (which is subject to corresponding duties under Schedule 2) or on the Secretary of State.

(6) In this section "accounting year" means, except so far as the Secretary of State otherwise directs, the period of twelve months ending with 31st March in any year.

60.—(1) It shall be the duty of the Secretary of State to secure that each Area Health Authority arranges for one of its officers who is a fully registered medical practitioner to furnish, on the application of an employment medical adviser, such particulars of the school medical record of a person who has not attained the age of eighteen and such other information relating to his medical history as the adviser may reasonably require for the efficient performance of his functions; but no particulars or information about any person which may be furnished to an adviser in pursuance of this subsection shall (without the consent of that person) be disclosed by the adviser otherwise than for the efficient performance of his functions. Supplementary.

(2) In its application to Scotland the preceding subsection shall have effect with the substitution of the words "every Health Board arrange for one of their" for the words from "each" to "its".

(3) The Secretary of State may by order made by statutory instrument subject to annulment in pursuance of a resolution of either House of Parliament modify the provisions of section 7(3) and (4) of the Employment and Training Act 1973 (which 1973 c. 50. require a person's period of continuous employment by a relevant body or in the civil service of the State to be treated, for the purposes of sections 1 and 2 of the Contracts of Employment Act 1972 and of certain provisions of the Industrial 1972 c. 53. Relations Act 1971 affecting the right of an employee not to 1971 c. 72. be unfairly dismissed, as increased by reference to previous

PART II periods of continous employment by such a body or in that service) for the purpose of securing that employment as an employment medical adviser by an authority other than the Secretary of State is similarly treated for those purposes.

An order under this subsection may be varied or revoked by a subsequent order thereunder.

(4) References to the chief employment medical adviser or a deputy chief employment medical adviser in any provision of an enactment or instrument made under an enactment shall be read as references to a person appointed for the purposes of that provision by the authority responsible for maintaining the employment medical advisory service.

1972 c. 28. (5) The following provisions of the Employment Medical Advisory Service Act 1972 (which are superseded by the preceding provisions of this Part or rendered unnecessary by provisions contained in Part I), namely sections 1 and 6 and Schedule 1, shall cease to have effect ; but—

(*a*) in so far as anything done under or by virtue of the said section 1 or Schedule 1 could have been done under or by virtue of a corresponding provision of Part I or this Part, it shall not be invalidated by the repeal of that section and Schedule by this Act but shall have effect as if done under or by virtue of that corresponding provision ; and

(*b*) any order made under the said section 6 which is in force immediately before the repeal of that section by this Act shall remain in force notwithstanding that repeal, but may be revoked or varied by regulations under section 43(2) or 57, as if it were an instrument containing regulations made under section 43(2) or 57, as the case may require.

(6) Where any Act (whether passed before, or in the same Session as, this Act) or any document refers, either expressly or by implication, to or to any enactment contained in any of the provisions of the said Act of 1972 which are mentioned in the preceding subsection, the reference shall, except where the context otherwise requires, be construed as, or as including, a reference to the corresponding provision of this Act.

1889 c. 63. (7) Nothing in subsection (5) or (6) above shall be taken as prejudicing the operation of section 38 of the Interpretation Act 1889 (which relates to the effect of repeals).

PART IV

MISCELLANEOUS AND GENERAL

Amendment of Radiological Protection Act 1970. 1970 c. 46.

77.—(1) Section 1 of the Radiological Protection Act 1970 (establishment and functions of the National Radiological Protection Board) shall be amended in accordance with the following provisions of this subsection—

(*a*) after subsection (6) there shall be inserted as subsection (6A)—

" (6A) In carrying out such of their functions as relate to matters to which the functions of the Health and Safety Commission relate, the Board shall (without prejudice to subsection (7) below) act in consultation with the Commission and have regard to the Commission's policies with respect to such matters." ;

(*b*) after subsection (7) there shall be inserted as subsections (7A) and (7B)—

" (7A) Without prejudice to subsection (6) or (7) above, it shall be the duty of the Board, if so directed by the Health Ministers, to enter into an agreement with the Health and Safety Commission for the Board to carry out on behalf of the Commission such of the Commission's functions relating to ionising or other radiations (including those which are not electro-magnetic) as may be determined by or in accordance with the direction; and the Board shall have power to carry out any agreement entered into in pursuance of a direction under this subsection.

(7B) The requirement as to consultation in subsection (7) above shall not apply to a direction under subsection 7(A)." ;

(*c*) in subsection (8), after the words " subsection (7) " there shall be inserted the words " or (7A) ".

(2) In section 2(6) of the Radiological Protection Act 1970 (persons by whom, as regards premises occupied by the said Board, sections 1 to 51 of the Offices, Shops and Railway Premises Act 1963 and regulations thereunder are enforceable) for the words from " inspectors appointed " to the end of the subsection there shall be substituted the words " inspectors appointed by the Health and Safety Executive under section 19 of the Health and Safety at Work etc. Act 1974."

1963 c. 41.

Amendment of Fire Precautions Act 1971. 1971 c. 40.

78.—(1) The Fire Precautions Act 1971 shall be amended in accordance with the following provisions of this section.

(2) In section 1(2) (power to designate uses of premises for which fire certificate is compulsory) at the end there shall be inserted as paragraph (*f*)—

" (*f*) use as a place of work."

PART IV

(3) In section 2 (premises exempt from section 1), paragraphs (*a*) to (*c*) (which exempt certain premises covered by the Offices, Shops and Railway Premises Act 1963, the Factories Act 1961 or the Mines and Quarries Act 1954) shall cease to have effect.

1963 c. 41.
1961 c. 34.
1954 c. 70.

(4) After section 9 there shall be inserted as section 9A—

" Duty to provide certain premises with means of escape in case of fire.

9A.—(1) All premises to which this section applies shall be provided with such means of escape in case of fire for the persons employed to work therein as may reasonably be required in the circumstances of the case.

(2) The premises to which this section applies are—

(*a*) office premises, shop premises and railway premises to which the Offices, Shops and Railway Premises Act 1963 applies; and

(*b*) premises which are deemed to be such premises for the purposes of that Act,

being (in each case) premises in which persons are employed to work.

(3) In determining, for the purposes of this section, what means of escape may reasonably be required in the case of any premises, regard shall be had (amongst other things) not only to the number of persons who may be expected to be working in the premises at any time but also to the number of persons (other than those employed to work therein) who may reasonably be expected to be resorting to the premises at that time.

(4) In the event of a contravention of subsection (1) above the occupier of the premises shall be guilty of an offence and liable on summary conviction to a fine not exceeding £400."

(5) In section 12(1) (power to make regulations about fire precautions as regards certain premises), at the end there shall be added the words " and nothing in this section shall confer on the Secretary of State power to make provision with respect to the taking or observance of special precautions in connection with the carrying on of any manufacturing process.

(6) In section 17 (duty of fire authorities to consult other authorities before requiring alterations to buildings)—

(*a*) in subsection (1), the word " and " shall be omitted where last occurring in paragraph (i) and shall be added at the end of paragraph (ii), and after paragraph

PART IV

(ii) there shall be added as paragraph (iii)—

"(iii) if the premises are used as a place of work and are within the field of responsibility of one or more enforcing authorities within the meaning of Part I of the Health and Safety at Work etc. Act 1974, consult that authority or each of those authorities.";

(*b*) in subsection (2) (clarification of references in section 9 to persons aggrieved), for the words "or buildings authority" there shall be substituted the words "buildings authority or other authority";

(*c*) after subsection (2) there shall be added as subsection (3)—

"(3) Section 18(7) of the Health and Safety at Work etc. Act 1974 (meaning in Part I of that Act of 'enforcing authority' and of such an authority's 'field of responsibility') shall apply for the purposes of this section as it applies for the purposes of that Part."

(7) In section 18 (enforcement of Act)—

(*a*) for the word "it" there shall be substituted the words "(1) Subject to subsection (2) below, it";

(*b*) for the word "section" there shall be substituted the word "subsection"; and

(*c*) after the word "offence" there shall be added as subsection (2)—

"(2) A fire authority shall have power to arrange with the Health and Safety Commission for such of the authority's functions under this Act as may be specified in the arrangements to be performed on their behalf by the Health and Safety Executive (with or without payment) in relation to any particular premises so specified which are used as a place of work."

(8) In section 40 (application to Crown etc.)—

(*a*) in subsection (1)(*a*) (provisions which apply to premises occupied by the Crown), after the word "6" there shall be inserted the words ", 9A (except subsection (4))";

(*b*) in subsection (1)(*b*) (provisions which apply to premises owned, but not occupied by, the Crown), after the word "8" there shall be inserted the word "9A";

(*c*) in subsection (10) (application of Act to hospital premises in Scotland), for the words from "Regional" to "hospitals" there shall be substituted the words "Health Board";

PART IV

(*d*) after subsection (10) there shall be inserted the following subsection—

"(10A) This Act shall apply to premises in England occupied by a Board of Governors of a teaching hospital (being a body for the time being specified in an order under section 15(1) of the National Health Service Reorganisation Act 1973) as if they were premises occupied by the Crown.". 1973 c. 12.

(9) In section 43(1) (interpretation) there shall be added at the end of the following definition—

"work" has the same meaning as it has for the purposes of Part I of the Health and Safety at Work etc. Act 1974".

(10) Schedule 8 (transitional provisions with respect to fire certificates under the Factories Act 1961 or the Offices, Shops and Railway Premises Act 1963) shall have effect. 1961 c. 34. 1963 c. 41.

Amendment of Companies Acts as to directors' reports. 1967 c. 81.

79.—(1) The Companies Act 1967 shall be amended in accordance with the following provisions of this section.

(2) In section 16 (additional general matters to be dealt with in directors' reports) in subsection (1) there shall be added after paragraph (*f*)—

"(*g*) in the case of companies of such classes as may be prescribed in regulations made by the Secretary of State, contain such information as may be so prescribed about the arrangements in force in that year for securing the health, safety and welfare at work of employees of the company and its subsidiaries and for protecting other persons against risks to health or safety arising out of or in connection with the activities at work of those employees."

(3) After subsection (4) of the said section 16 there shall be added—

"(5) Regulations made under paragraph (*g*) of subsection (1) above may—

(*a*) make different provision in relation to companies of different classes;

(*b*) enable any requirements of the regulations to be dispensed with or modified in particular cases by any specified person or by any person authorised in that behalf by a specified authority;

(*c*) contain such transitional provisions as the Secretary of State thinks necessary or expedient in connection with any provision made by the regulations.

PART IV

(6) The power to make regulations under the said paragraph (*g*) shall be exercisable by statutory instrument which shall be subject to annulment in pursuance of a resolution of either House of Parliament.

(7) Any expression used in the said paragraph (*g*) and in Part I of the Health and Safety at Work etc. Act 1974 shall have the same meaning in that paragraph as it has in that Part of that Act and section 1(3) of that Act shall apply for interpreting that paragraph as it applies for interpreting that Part of that Act; and in subsection (5) above "specified" means specified in regulations made under that paragraph.".

General power to repeal or modify Acts and instruments.

80.—(1) Regulations made under this subsection may repeal or modify any provision to which this subsection applies if it appears to the authority making the regulations that the repeal or, as the case may be, the modification of that provision is expedient in consequence of or in connection with any provision made by or under Part I.

(2) Subsection (1) above applies to any provision, not being among the relevant statutory provisions, which—

(*a*) is contained in this Act or in any other Act passed before or in the same Session as this Act; or

(*b*) is contained in any regulations, order or other instrument of a legislative character which was made under an Act before the passing of this Act; or

(*c*) applies, excludes or for any other purpose refers to any of the relevant statutory provisions and is contained in any regulations, order or other instrument of a legislative character which is made under an Act but does not fall within paragraph (*b*) above.

(3) Without prejudice to the generality of subsection (1) above, the modifications which may be made by regulations thereunder include modifications relating to the enforcement of provisions to which this section applies (including the appointment of persons for the purpose of such enforcement, and the powers of persons so appointed).

(4) The power to make regulations under subsection (1) above shall be exercisable—

(*a*) in relation to provisions not relating exclusively to agricultural operations, by the Secretary of State;

(*b*) in relation to provisions relating exclusively to the relevant agricultural purposes, by the appropriate agriculture authority;

but before making regulations under that subsection the Secretary of State or the appropriate agriculture authority shall consult such bodies as appear to the Secretary of State or, as the case may be, that authority to be appropriate. PART IV

(5) Regulations under subsection (1) above in relation to provisions falling within subsection (4)(*b*) above may be either regulations applying to Great Britain and made by the Minister of Agriculture, Fisheries and Food and the Secretary of State acting jointly, or regulations applying to England and Wales only and made by the said Minister, or regulations applying to Scotland only and made by the Secretary of State; and in subsection (4)(*b*) above " the appropriate agriculture authority " shall be construed accordingly.

(6) In this section " the relevant statutory provisions," " the relevant agricultural purposes " and " agricultural operation " have the same meaning as in Part I.

Expenses and receipts.

81. There shall be paid out of money provided by Parliament—

(*a*) any expenses incurred by a Minister of the Crown or government department for the purposes of this Act; and

(*b*) any increase attributable to the provisions of this Act in the sums payable under any other Act out of money so provided;

and any sums received by a Minister of the Crown or government department by virtue of this Act shall be paid into the Consolidated Fund.

General provisions as to interpretation and regulations.

82.—(1) In this Act—

(*a*) " Act " includes a provisional order confirmed by an Act;

(*b*) " contravention " includes failure to comply, and " contravene " has a corresponding meaning;

(*c*) " modifications " includes additions, omissions and amendments, and related expressions shall be construed accordingly;

(*d*) any reference to a Part, section or Schedule not otherwise identified is a reference to that Part or section of, or Schedule to, this Act.

(2) Except in so far as the context otherwise requires, any reference in this Act to an enactment is a reference to it as amended, and includes a reference to it as applied, by or under any other enactment, including this Act.

PART IV

(3) Any power conferred by Part I or II or this Part to make regulations—

(*a*) includes power to make different provision by the regulations for different circumstances or cases and to include in the regulations such incidental, supplemental and transitional provisions as the authority making the regulations considers appropriate in connection with the regulations; and

(*b*) shall be exercisable by statutory instrument, which shall be subject to annulment in pursuance of a resolution of either House of Parliament.

Minor and consequential amendments, and repeals.

83.—(1) The enactments mentioned in Schedule 9 shall have effect subject to the amendments specified in that Schedule (being minor amendments or amendments consequential upon the provisions of this Act).

(2) The enactments mentioned in Schedule 10 are hereby repealed to the extent specified in the third column of that Schedule.

Extent, and application of Act.

84.—(1) This Act, except—

(*a*) Part I and this Part so far as may be necessary to enable regulations under section 15 or 30 to be made and operate for the purpose mentioned in paragraph 2 of Schedule 3; and

(*b*) paragraphs 2 and 3 of Schedule 9.

does not extend to Northern Ireland.

(2) Part III, except section 75 and Schedule 7, does not extend to Scotland.

(3) Her Majesty may by Order in Council provide that the provisions of Parts I and II and this Part shall, to such extent and for such purposes as may be specified in the Order, apply (with or without modification) to or in relation to persons, premises, work, articles, substances and other matters (of whatever kind) outside Great Britain as those provisions apply within Great Britain or within a part of Great Britain so specified.

For the purposes of this subsection "premises", "work" and "substance" have the same meaning as they have for the purposes of Part I.

(4) An Order in Council under subsection (3) above—

(*a*) may make different provision for different circumstances or cases;

(*b*) may (notwithstanding that this may affect individuals or bodies corporate outside the United Kingdom) provide for any of the provisions mentioned in that

PART IV

subsection, as applied by such an Order, to apply to individuals whether or not they are British subjects and to bodies corporate whether or not they are incorporated under the law of any part of the United Kingdom;

(*c*) may make provision for conferring jurisdiction on any court or class of courts specified in the Order with respect to offences under Part I committed outside Great Britain or with respect to causes of action arising by virtue of section 47(2) in respect of acts or omissions taking place outside Great Britain, and for the determination, in accordance with the law in force in such part of Great Britain as may be specified in the Order, of questions arising out of such acts or omissions;

(*d*) may exclude from the operation of section 3 of the Territorial Waters Jurisdiction Act 1878 (consents required for prosecutions) proceedings for offences under any provision of Part I committed outside Great Britain; 1878 c. 73.

(*e*) may be varied or revoked by a subsequent Order in Council under this section;

and any such Order shall be subject to annulment in pursuance of a resolution of either House of Parliament.

(5) In relation to proceedings for an offence under Part I committed outside Great Britain by virtue of an Order in Council under subsection (3) above, section 38 shall have effect as if the words " by an inspector, or " were omitted.

(6) Any jurisdiction conferred on any court under this section shall be without prejudice to any jurisdiction exercisable apart from this section by that or any other court.

Short title and commencement.

85.—(1) This Act may be cited as the Health and Safety at Work etc. Act 1974.

(2) This Act shall come into operation on such day as the Secretary of State may by order made by statutory instrument appoint, and different days may be appointed under this subsection for different purposes.

(3) An order under this section may contain such transitional provisions and savings as appear to the Secretary of State to be necessary or expedient in connection with the provisions thereby brought into force, including such adaptations of those provisions or any provision of this Act then in force as appear to him to be necessary or expedient in consequence of the partial operation of this Act (whether before or after the day appointed by the order).

SCHEDULES

Sections 1 and 53.

SCHEDULE 1

Existing Enactments which are Relevant Statutory Provisions

Chapter	Short title	Provisions which are relevant statutory provisions
1875 c. 17.	The Explosives Act 1875.	The whole Act except sections 30 to 32, 80 and 116 to 121.
1882 c. 22.	The Boiler Explosions Act 1882.	The whole Act.
1890 c. 35.	The Boiler Explosions Act 1890.	The whole Act.
1906 c. 14.	The Alkali, &c. Works Regulation Act 1906.	The whole Act.
1909 c. 43.	The Revenue Act 1909.	Section 11.
1919 c. 23.	The Anthrax Prevention Act 1919.	The whole Act.
1920 c. 65.	The Employment of Women, Young Persons and Children Act 1920.	The whole Act.
1922 c. 35.	The Celluloid and Cinematograph Film Act 1922.	The whole Act.
1923 c. 17.	The Explosives Act 1923.	The whole Act.
1926 c. 43.	The Public Health (Smoke Abatement) Act 1926.	The whole Act.
1928 c. 32.	The Petroleum (Consolidation) Act 1928.	The whole Act.
1936 c. 22.	The Hours of Employment (Conventions) Act 1936.	The whole Act except section 5.
1936 c. 27.	The Petroleum (Transfer of Licences) Act 1936.	The whole Act.
1937 c. 45.	The Hydrogen Cyanide (Fumigation) Act 1937.	The whole Act.
1945 c. 19.	The Ministry of Fuel and Power Act 1945.	Section 1(1) so far as it relates to maintaining and improving the safety, health and welfare of persons employed in or about mines and quarries in Great Britain.
1946 c. 59.	The Coal Industry Nationalisation Act 1946.	Section 42(1) and (2).
1948 c. 37.	The Radioactive Substances Act 1948.	Section 5(1)(*a*).
1951 c. 21.	The Alkali, &c. Works Regulation (Scotland) Act 1951.	The whole Act.
1951 c. 58.	The Fireworks Act 1951.	Sections 4 and 7.
1952 c. 60.	The Agriculture (Poisonous Substances) Act 1952.	The whole Act.

Sch. 1

Chapter	Short title	Provisions which are relevant statutory provisions
1953 c. 47.	The Emergency Laws (Miscellaneous Provisions) Act 1953.	Section 3.
1954 c. 70.	The Mines and Quarries Act 1954.	The whole Act except section 151.
1956 c. 49.	The Agriculture (Safety, Health and Welfare Provisions) Act 1956.	The whole Act.
1961 c. 34.	The Factories Act 1961.	The whole Act except section 135.
1961 c. 64.	The Public Health Act 1961.	Section 73.
1962 c. 58.	The Pipe-lines Act 1962.	Sections 20 to 26, 33, 34 and 42, Schedule 5.
1963 c. 41.	The Offices, Shops and Railway Premises Act 1963.	The whole Act.
1965 c. 57.	The Nuclear Installations Act 1965.	Sections 1, 3 to 6, 22 and 24, Schedule 2.
1969 c. 10.	The Mines and Quarries (Tips) Act 1969.	Sections 1 to 10.
1971 c. 20.	The Mines Management Act 1971.	The whole Act.
1972 c. 28.	The Employment Medical Advisory Service Act 1972.	The whole Act except sections 1 and 6 and Schedule 1.

SCHEDULE 2

Section 10.

Additional Provisions Relating to Constitution etc. of the Commission and Executive

Tenure of office

1. Subject to paragraphs 2 to 4 below, a person shall hold and vacate office as a member or as chairman or deputy chairman in accordance with the terms of the instrument appointing him to that office.

2. A person may at any time resign his office as a member or as chairman or deputy chairman by giving the Secretary of State a notice in writing signed by that person and stating that he resigns that office.

3.—(1) If a member becomes or ceases to be the chairman or deputy chairman, the Secretary of State may vary the terms of the instrument appointing him to be a member so as to alter the date on which he is to vacate office as a member.

(2) If the chairman or deputy chairman ceases to be a member he shall cease to be chairman or deputy chairman, as the case may be.

SCH. 2

4.—(1) If the Secretary of State is satisfied that a member—

(*a*) has been absent from meetings of the Commission for a period longer than six consecutive months without the permission of the Commission ; or

(*b*) has become bankrupt or made an arrangement with his creditors ; or

(*c*) is incapacitated by physical or mental illness ; or

(*d*) is otherwise unable or unfit to discharge the functions of a member,

the Secretary of State may declare his office as a member to be vacant and shall notify the declaration in such manner as the Secretary of State thinks fit ; and thereupon the office shall become vacant.

(2) In the application of the preceding sub-paragraph to Scotland for the references in paragraph (*b*) to a member's having become bankrupt and to a member's having made an arrangement with his creditors there shall be substituted respectively references to sequestration of a member's estate having been awarded and to a member's having made a trust deed for behoof of his creditors or a composition contract.

Remuneration etc. of members

5. The Commission may pay to each member such remuneration and allowances as the Secretary of State may determine.

6. The Commission may pay or make provision for paying, to or in respect of any member, such sums by way of pension, superannuation allowances and gratuities as the Secretary of State may determine.

7. Where a person ceases to be a member otherwise than on the expiry of his term of office and it appears to the Secretary of State that there are special circumstances which make it right for him to receive compensation, the Commission may make to him a payment of such amount as the Secretary of State may determine.

Proceedings

8. The quorum of the Commission and the arrangements relating to meetings of the Commission shall be such as the Commission may determine.

9. The validity of any proceedings of the Commission shall not be affected by any vacancy among the members or by any defect in the appointment of a member.

Staff

10. It shall be the duty of the Executive to provide for the Commission such officers and servants as are requisite for the proper discharge of the Commission's functions ; and any reference in this Act to an officer or servant of the Commission is a reference to an officer or servant provided for the Commission in pursuance of this paragraph.

11. The Executive may appoint such officers and servants as it may determine with the consent of the Secretary of State as to numbers and terms and conditions of service. SCH. 2

12. The Commission shall pay to the Minister for the Civil Service, at such times in each accounting year as may be determined by that Minister subject to any directions of the Treasury, sums of such amounts as he may so determine for the purposes of this paragraph as being equivalent to the increase during that year of such liabilities of his as are attributable to the provision of pensions, allowances or gratuities to or in respect of persons who are or have been in the service of the Executive in so far as that increase results from the service of those persons during that accounting year and to the expense to be incurred in administering those pensions, allowances or gratuities.

Performance of functions

13. The Commission may authorise any member of the Commission or any officer or servant of the Commission or of the Executive to perform on behalf of the Commission such of the Commission's functions (including the function conferred on the Commission by this paragraph) as are specified in the authorisation.

Accounts and reports

14.—(1) It shall be the duty of the Commission—

(*a*) to keep proper accounts and proper records in relation to the accounts ;

(*b*) to prepare in respect of each accounting year a statement of accounts in such form as the Secretary of State may direct with the approval of the Treasury ; and

(*c*) to send copies of the statement to the Secretary of State and the Comptroller and Auditor General before the end of the month of November next following the accounting year to which the statement relates.

(2) The Comptroller and Auditor General shall examine, certify and report on each statement received by him in pursuance of this Schedule and shall lay copies of each statement and of his report before each House of Parliament.

15. It shall be the duty of the Commission to make to the Secretary of State, as soon as possible after the end of each accounting year, a report on the performance of its functions during that year ; and the Secretary of State shall lay before each House of Parliament a copy of each report made to him in pursuance of this paragraph.

Supplemental

16. The Secretary of State shall not make a determination or give his consent in pursuance of paragraph 5, 6, 7 or 11 of this Schedule except with the approval of the Minister for the Civil Service.

Sch. 2

17. The fixing of the common seal of the Commission shall be authenticated by the signature of the secretary of the Commission or some other person authorised by the Commission to act for that purpose.

18. A document purporting to be duly executed under the seal of the Commission shall be received in evidence and shall, unless the contrary is proved, be deemed to be so executed.

19. In the preceding provisions of this Schedule—

(*a*) " accounting year " means the period of twelve months ending with 31st March in any year except that the first accounting year of the Commission shall, if the Secretary of State so directs, be such period shorter or longer than twelve months (but not longer than two years) as is specified in the direction ; and

(*b*) " the chairman ", " a deputy chairman " and " a member " mean respectively the chairman, a deputy chairman and a member of the Commission.

20.—(1) The preceding provisions of this Schedule (except paragraphs 10 to 12 and 15) shall have effect in relation to the Executive as if—

(*a*) for any reference to the Commission there were substituted a reference to the Executive ;

(*b*) for any reference to the Secretary of State in paragraphs 2 to 4 and 19 and the first such reference in paragraph 7 there were substituted a reference to the Commission ;

(*c*) for any reference to the Secretary of State in paragraphs 5 to 7 (except the first such reference in paragraph 7) there were substituted a reference to the Commission acting with the consent of the Secretary of State ;

(*d*) for any reference to the chairman there were substituted a reference to the director, and any reference to the deputy chairman were omitted ;

(*e*) in paragraph 14(1)(*c*) for the words from " Secretary " to " following " there were substituted the words " Commission by such date as the Commission may direct after the end of ".

(2) It shall be the duty of the Commission to include in or send with the copies of the statement sent by it as required by paragraph 14(1)(*c*) of this Schedule copies of the statement sent to it by the Executive in pursuance of the said paragraph 14(1)(*c*) as adapted by the preceding sub-paragraph.

(3) The terms of an instrument appointing a person to be a member of the Executive shall be such as the Commission may determine with the approval of the Secretary of State and the Minister for the Civil Service.

SCHEDULE 3

Section 15.

Subject-Matter of Health and Safety Regulations

1.—(1) Regulating or prohibiting—

(*a*) the manufacture, supply or use of any plant;

(*b*) the manufacture, supply, keeping or use of any substance;

(*c*) the carrying on of any process or the carrying out of any operation.

(2) Imposing requirements with respect to the design, construction, guarding, siting, installation, commissioning, examination, repair, maintenance, alteration, adjustment, dismantling, testing or inspection of any plant.

(3) Imposing requirements with respect to the marking of any plant or of any articles used or designed for use as components in any plant, and in that connection regulating or restricting the use of specified markings.

(4) Imposing requirements with respect to the testing, labelling or examination of any substance.

(5) Imposing requirements with respect to the carrying out of research in connection with any activity mentioned in sub-paragraphs (1) to (4) above.

2.—(1) Prohibiting the importation into the United Kingdom or the landing or unloading there of articles or substances of any specified description, whether absolutely or unless conditions imposed by or under the regulations are complied with.

(2) Specifying, in a case where an act or omission in relation to such an importation, landing or unloading as is mentioned in the preceding sub-paragraph constitutes an offence under a provision of this Act and of the Customs and Excise Act 1952, the Act under which the offence is to be punished. 1952 c. 44.

3.—(1) Prohibiting or regulating the transport of articles or substances of any specified description.

(2) Imposing requirements with respect to the manner and means of transporting articles or substances of any specified description, including requirements with respect to the construction, testing and marking of containers and means of transport and the packaging and labelling of articles or substances in connection with their transport.

4.—(1) Prohibiting the carrying on of any specified activity or the doing of any specified thing except under the authority and in accordance with the terms and conditions of a licence, or except with the consent or approval of a specified authority.

(2) Providing for the grant, renewal, variation, transfer and revocation of licences (including the variation and revocation of conditions attached to licences).

Sch. 3 5. Requiring any person, premises or thing to be registered in any specified circumstances or as a condition of the carrying on of any specified activity or the doing of any specified thing.

6.—(1) Requiring, in specified circumstances, the appointment (whether in a specified capacity or not) of persons (or persons with specified qualifications or experience, or both) to perform specified functions, and imposing duties or conferring powers on persons appointed (whether in pursuance of the regulations or not) to perform specified functions.

(2) Restricting the performance of specified functions to persons possessing specified qualifications or experience.

7. Regulating or prohibiting the employment in specified circumstances of all persons or any class of persons.

8.—(1) Requiring the making of arrangements for securing the health of persons at work or other persons, including arrangements for medical examinations and health surveys.

(2) Requiring the making of arrangements for monitoring the atmospheric or other conditions in which persons work.

9. Imposing requirements with respect to any matter affecting the conditions in which persons work, including in particular such matters as the structural condition and stability of premises, the means of access to and egress from premises, cleanliness, temperature, lighting, ventilation, overcrowding, noise, vibrations, ionising and other radiations, dust and fumes.

10. Securing the provision of specified welfare facilities for persons at work, including in particular such things as an adequate water supply, sanitary conveniences, washing and bathing facilities, ambulance and first-aid arrangements, cloakroom accommodation, sitting facilities and refreshment facilities.

11. Imposing requirements with respect to the provision and use in specified circumstances of protective clothing or equipment, including clothing affording protection against the weather.

12. Requiring in specified circumstances the taking of specified precautions in connection with the risk of fire.

13.—(1) Prohibiting or imposing requirements in connection with the emission into the atmosphere of any specified gas, smoke or dust or any other specified substance whatsoever.

(2) Prohibiting or imposing requirements in connection with the emission of noise, vibrations or any ionising or other radiations.

(3) Imposing requirements with respect to the monitoring of any such emission as is mentioned in the preceding sub-paragraphs.

14. Imposing requirements with respect to the instruction, training and supervision of persons at work. SCH. 3

15.—(1) Requiring, in specified circumstances, specified matters to be notified in a specified manner to specified persons.

(2) Empowering inspectors in specified circumstances to require persons to submit written particulars of measures proposed to be taken to achieve compliance with any of the relevant statutory provisions.

16. Imposing requirements with respect to the keeping and preservation of records and other documents, including plans and maps.

17. Imposing requirements with respect to the management of animals.

18. The following purposes as regards premises of any specified description where persons work, namely—

(*a*) requiring precautions to be taken against dangers to which the premises or persons therein are or may be exposed by reason of conditions (including natural conditions) existing in the vicinity ;

(*b*) securing that persons in the premises leave them in specified circumstances.

19. Conferring, in specified circumstances involving a risk of fire or explosion, power to search a person or any article which a person has with him for the purpose of ascertaining whether he has in his possession any article of a specified kind likely in those circumstances to cause a fire or explosion, and power to seize and dispose of any article of that kind found on such a search.

20. Restricting, prohibiting or requiring the doing of any specified thing where any accident or other occurrence of a specified kind has occurred.

21. As regards cases of any specified class, being a class such that the variety in the circumstances of particular cases within it calls for the making of special provision for particular cases, any of the following purposes, namely—

(*a*) conferring on employers or other persons power to make rules or give directions with respect to matters affecting health or safety ;

(*b*) requiring employers or other persons to make rules with respect to any such matters ;

(*c*) empowering specified persons to require employers or other persons either to make rules with respect to any such matters or to modify any such rules previously made by virtue of this paragraph ; and

SCH. 3

(*d*) making admissible in evidence without further proof, in such circumstances and subject to such conditions as may be specified, documents which purport to be copies of rules or rules of any specified class made under this paragraph.

22. Conferring on any local or public authority power to make byelaws with respect to any specified matter, specifying the authority or person by whom any byelaws made in the exercise of that power need to be confirmed, and generally providing for the procedure to be followed in connection with the making of any such byelaws.

Interpretation

23.—(1) In this Schedule " specified " means specified in health and safety regulations.

(2) It is hereby declared that the mention in this Schedule of a purpose that falls within any more general purpose mentioned therein is without prejudice to the generality of the more general purpose.

Section 32.

SCHEDULE 4

MODIFICATIONS OF PART I IN CONNECTION WITH AGRICULTURE

Provisions applied	*Modifications*
1. Section 13(1) (various powers).	(*a*) Paragraph (*b*) shall be omitted; (*b*) references to the Commission or the Secretary of State shall be read as references to the appropriate Agriculture Minister, so however that references to the Commission's functions shall be read as references to the functions of that Minister under the relevant statutory provisions in relation to matters relating exclusively to the relevant agricultural purposes.
2. Section 14 (power to direct investigations and inquiries).	(*a*) References to the Commission shall be read as references to the appropriate Agriculture Minister; (*b*) in subsection (1), the reference to the general purposes of Part I shall be read as a reference to the relevant agricultural purposes; (*c*) in subsection (2), for the words from " direct " to " other " in paragraph (*a*) there shall be substituted the words " authorise any ", the words " with the consent of the Secretary of State " shall be omitted, and for the words from " only matters " to the end of the subsection there shall be substituted the words " matters relating exclusively to the relevant agricultural purposes ";

Provisions applied	*Modifications*
	(*d*) in subsection (6), references to the Secretary of State shall be read as references to the appropriate Agriculture Minister.
3. Section 16 (approval of codes of practice).	(*a*) In subsection (1), the reference to health and safety regulations shall be read as a reference to agricultural health and safety regulations and the words from " and except " to " agricultural operations " shall be omitted, but so that the section shall confer power to approve or issue codes of practice for any provision mentioned in section 16(1) only for the purposes of the application of that provision to matters relating exclusively to the relevant agricultural purposes;
	(*b*) a code of practice may either be approved for Great Britain and be so approved by the Minister of Agriculture, Fisheries and Food and the Secretary of State acting jointly, or be approved for England and Wales only and be so approved by that Minister or be approved for Scotland only and be so approved by the Secretary of State, and the references to the Commission shall accordingly be read as references to the Agriculture Ministers or the said Minister or the Secretary of State as the case may require;
	(*c*) for subsection (2) there shall be substituted— " (2) Before approving a code of practice under subsection (1) above the Minister or Ministers proposing to do so shall consult the Commission and any other body that appears to him or them to be appropriate.";
	(*d*) for subsection (5) there shall be substituted— " (5) The authority by whom a code of practice has been approved under this section may at any time withdraw approval from that code, but before doing so shall consult the same bodies as the authority would be required to consult under subsection (2) above if the authority were proposing to approve the code.".

SCH. 4

Provisions applied	*Modifications*
4. Section 17(3) (use of approved codes in criminal proceedings).	The reference to the Commission shall be read as a reference to the Agriculture Ministers or either of them.
5. Section 27 (obtaining of information).	(*a*) References to the Commission or the Executive shall be read as references to the appropriate Agriculture Minister, so however that references to the Commission's functions shall be read as references to the functions of that Minister under the relevant statutory provisions in relation to matters relating exclusively to the relevant agricultural purposes; (*b*) references to an enforcing authority's functions shall be read as references to an enforcing authority's functions under the relevant statutory provisions in relation to matters relating exclusively to the relevant agricultural purposes; (*c*) in subsection (1), the words " with the consent of the Secretary of State " shall be omitted; (*d*) in subsection (2)(*b*), the reference to the Secretary of State shall be read as a reference to the appropriate Agriculture Minister, and the words " and the recipient of the information " shall be omitted.

Section 61.

SCHEDULE 5

SUBJECT-MATTER OF BUILDING REGULATIONS

1. Preparation of sites.

2. Suitability, durability and use of materials and components (including surface finishes).

3. Structural strength and stability, including—

(*a*) precautions against overloading, impact and explosion ;

(*b*) measures to safeguard adjacent buildings and services ;

(*c*) underpinning.

4. Fire precautions, including—

(*a*) structural measures to resist the outbreak and spread of fire and to mitigate its effects ;

(*b*) services, fittings and equipment designed to mitigate the effects of fire or to facilitate fire-fighting ;

(*c*) means of escape in case of fire and means for securing that such means of escape can be safely and effectively used at all material times.

5. Resistance to moisture and decay. Sch. 5

6. Measures affecting the transmission of heat.

7. Measures affecting the transmission of sound.

8. Measures to prevent infestation.

9. Measures affecting the emission of smoke, gases, fumes, grit or dust or other noxious or offensive substances.

10. Drainage (including waste disposal units).

11. Cesspools and other means for the reception, treatment or disposal of foul matter.

12. Storage, treatment and removal of waste.

13. Installations utilising solid fuel, oil, gas, electricity or any other fuel or power (including appliances, storage tanks, heat exchangers, ducts, fans and other equipment).

14. Water services (including wells and bore-holes for the supply of water) and fittings and fixed equipment associated therewith.

15. Telecommunications services (including telephones and radio and television wiring installations).

16. Lifts, escalators, hoists, conveyors and moving footways.

17. Plant providing air under pressure.

18. Standards of heating, artificial lighting, mechanical ventilation and air-conditioning and provision of power outlets.

19. Open space about buildings and the natural lighting and ventilation of buildings.

20. Accommodation for specific purposes in or in connection with buildings, and the dimensions of rooms and other spaces within buildings.

21. Means of access to and egress from buildings and parts of buildings.

22. Prevention of danger and obstruction to persons in and about buildings (including passers-by).

23. Matters connected with or ancillary to any of the matters mentioned in the preceding provisions of this Schedule.

SCHEDULE 6

Section 61.

Amendments of Enactments Relating to Building Regulations

Part I

Amendments

Amendments of Public Health Act 1936 — 1936 c. 49.

1. In section 64 of the 1936 Act (passing or rejection of plans)—

(*a*) for subsection (3) substitute—

"(3) Where plans of any proposed work deposited with a local authority are rejected in pursuance of the preceding provisions of this section, the person by whom or

SCH. 6

on whose behalf they were deposited may appeal against the rejection to the Secretary of State within the prescribed time and in the prescribed manner; and where the rejection results wholly or partly from the fact that a person or body whose approval or satisfaction in any respect is required by the regulations has withheld approval or not been satisfied, an appeal under this subsection may be brought on (or on grounds which include) the ground that the person or body in question ought in the circumstances to have approved or been satisfied in that respect." ; and

(*b*) subsection (4) shall cease to have effect.

2. In section 65 of the 1936 Act (power to require removal or alteration of work not in conformity with building regulations etc.)—

(*a*) in subsection (1), after " therein " insert " and additions thereto and to execute such additional work in connection therewith " ;

(*b*) after subsection (2) insert as subsection (2A)—

" (2A) Where a local authority have power to serve a notice under subsection (1) or (2) above on the owner of any work, they may in addition or instead serve such a notice on one or more of the following persons, namely the occupier and any builder or other person appearing to the authority to have control over the work." ;

(*c*) in subsection (3), after " therein " insert " and additions thereto and execute such additional work in connection therewith ", and at the end add as a proviso—

" Provided that where a notice under subsection (1) or (2) above is given to two or more persons in pursuance of subsection (2A) above, then—

(*a*) if they are given the notices on different dates, the said period of twenty-eight days shall for each of them run from the later or latest of those dates ; and

(*b*) if the notice is not complied with before the expiration of the said period or such longer period as a court of summary jurisdiction may on the application of any of them allow, any expenses recoverable as aforesaid may be recovered from any of them." ; and

(*d*) in subsection (4), for " or subsection (2) " substitute " , (2) or (2A) ", and at the end add as a proviso—

" Provided that, in a case where plans were deposited nothing in this subsection shall be taken to prevent such a notice from being given (before the expiration of twelve months from the completion of the work in question) in respect of anything of which particulars were not required to be shown in the plans.".

3. In section 90 of the 1936 Act (interpretation of Part II of that Act)—

(*a*) in subsection (2) (extended meaning, in that Part and building regulations, of references to the erection of a building), for the words from " and, so far " to " those regulations " substitute " except sections 61 to 71 and any other enactment to which section 74(1) of the Health and Safety at Work etc. Act 1974 applies " ; and

(*b*) for subsection (3) (meaning of references to deposited plans) substitute—

" (3) In this Part of this Act, unless the context otherwise requires,—

(*a*) any reference to the deposit of plans in accordance with building regulations shall be construed as a reference to the deposit of plans in accordance with those regulations for the purposes of section 64 of this Act ; and

(*b*) " plans " includes drawings of any other description and also specifications or other information in any form, and any reference to the deposit of plans shall be construed accordingly."

Amendments of Public Health Act 1961 1961 c. 64.

4. In section 4 of the 1961 Act (power to make building regulations)—

(*a*) in subsection (2) (power to make different provision for different areas) at the end add " and generally different provision for different circumstances or cases " ; and

(*b*) in subsection (6) (penalties for contravening building regulations) after " building regulations " insert " other than a provision designated in the regulations as one to which this subsection does not apply,", and for " one hundred pounds " and " ten pounds " substitute respectively " £400 " and " £50 ".

5. In section 6 of the 1961 Act (power to dispense with or relax requirements of building regulations)—

(*a*) in subsection (1), add at the end the words " either unconditionally or subject to compliance with any conditions specified in the direction, being conditions with respect to matters directly connected with the dispensation or relaxation." ;

(*b*) in the proviso to subsection (2), for the words from " shall " onwards substitute " may except applications of any description " ;

(*c*) for subsection (6) substitute—

" (6) An application by a local authority in connection with a building or proposed building in the area of that authority shall be made to the Secretary of State except where the power of giving the direction is exercisable by that authority." ;

SCH. 6

(*d*) after subsection (7), there shall be inserted as subsections (7A) and (7B)—

" (7A) If, on an application to the Secretary of State for a direction under this section, the Secretary of State considers that any requirement of building regulations to which the application relates is not applicable or is not or would not be contravened in the case of the work or proposed work to which the application relates, he may so determine and may give any directions that he considers necessary in the circumstances.

(7B) A person who contravenes any condition specified in a direction given under this section or permits any such condition to be contravened shall be liable to a fine not exceeding £400 and to a further fine not exceeding £50 for each day on which the offence continues after he is convicted." ; and

(*e*) subsection (8) shall be omitted.

6. In section 7 of the 1961 Act (appeal against local authority's refusal to dispense with or relax requirements of building regulations)—

(*a*) in subsection (1), after second " relax " insert " or grant such an application subject to conditions ", for " by notice in writing " substitute " in the prescribed manner ", for " one month " substitute " the prescribed period " and for " refusal " substitute " decision on the application " ;

(*b*) in subsection (2), for the words from " a period " to " and the local authority " substitute " the prescribed period " ;

(*c*) subsections (3) to (6) shall be omitted ; and

(*d*) at the end there shall be added the following subsection:—

" (7) Section 6(7A) of this Act shall apply in relation to an appeal to the Secretary of State under this section as it applies in relation to an application to him for a direction under section 6.".

7. For section 8 of the 1961 Act (advertisement of proposal to relax building regulations) substitute—

" Opportunity to make representations about proposal to relax building regulations.

8.—(1) Before the Secretary of State or a local authority give a direction under section 6 of this Act the prescribed steps shall be taken for affording to persons likely to be affected by the direction an opportunity to make representations about it ; and before giving the direction the Secretary of State or, as the case may be, the local authority shall consider any representations duly made in accordance with the regulations.

(2) Building regulations—

(*a*) may make provision as to the time to be allowed for making representations under the preceding subsection ;

(*b*) may require an applicant for such a direction, as a condition that his application shall be entertained, to pay or undertake to pay the cost of publishing any notice which is required by the regulations to be published in connection with the application ; and Sch. 6

(*c*) may exclude the requirements of the preceding subsection in prescribed cases.".

8. In section 9(3) of the 1961 Act (consultation with Building Regulations Advisory Committee and other bodies before making building regulations), at the end add " (including in particular, as regards regulations relevant to any of their functions, the National Water Council).".

Part II

Public Health Act 1936 Section 65 and Public Health Act 1961 Sections 4, 6 and 7 as amended

1936 c. 49.
1961 c. 64.

The Public Health Act 1936

65.—(1) If any work to which building regulations are applicable contravenes any of those regulations, the authority, without prejudice to their right to take proceedings for a fine in respect of the contravention, may by notice require the owner either to pull down or remove the work or, if he so elects, to effect such alterations therein and additions thereto and to execute such additional work in connection therewith as may be necessary to make it comply with the regulations.

(2) If, in a case where the local authority are by any section of this Act other than the last preceding section expressly required or authorised to reject plans, any work to which building regulations are applicable is executed either without plans having been deposited, or notwithstanding the rejection of the plans, or otherwise than in accordance with any requirements subject to which the authority passed the plans, the authority may by notice to the owner either require him to pull down or remove the work, or require him either to pull down or remove the work or, if he so elects, to comply with any other requirements specified in the notice, being requirements which they might have made under the section in question as a condition of passing plans.

(2A) Where a local authority have power to serve a notice under subsection (1) or (2) above on the owner of any work, they may in addition or instead serve such a notice on one or more of the following persons, namely the occupier and any builder or other person appearing to the authority to have control over the work.

(3) If a person to whom a notice has been given under the foregoing provisions of this section fails to comply with the notice before the expiration of twenty-eight days, or such longer period as a court

Sch. 6

of summary jurisdiction may on his application allow, the local authority may pull down or remove the work in question, or effect such alterations therein and additions thereto and execute such additional work in connection therewith as they deem necessary, and may recover from him the expenses reasonably incurred by them in so doing:

Provided that where a notice under subsection (1) or (2) above is given to two or more persons in pursuance of subsection (2A) above, then—

(*a*) if they are given the notices on different dates, the said period of twenty-eight days shall for each of them run from the later or latest of those dates ; and

(*b*) if the notice is not complied with before the expiration of the said period or such longer period as a court of summary jurisdiction may on the application of any of them allow, any expenses recoverable as aforesaid may be recovered from any of them.

(4) No such notice as is mentioned in subsection (1), (2) or (2A) of this section shall be given after the expiration of twelve months from the date of the completion of the work in question, and, in any case where plans were deposited, it shall not be open to the authority to give such a notice on the ground that the work contravenes any building regulation or, as the case may be, does not comply with their requirements under any such section of this Act as aforesaid, if either the plans were passed by the authority, or notice of their rejection was not given within the prescribed period from the deposit thereof, and if the work has been executed in accordance with the plans and of any requirement made by the local authority as a condition of passing the plans:

Provided that, in a case where plans were deposited, nothing in this subsection shall be taken to prevent such a notice from being given (before the expiration of twelve months from the completion of the work in question) in respect of anything of which particulars were not required to be shown in the plans.

(5) Nothing in this section shall affect the right of a local authority, or of the Attorney General, or any other person, to apply for an injunction for the removal or alteration of any work on the ground that it contravenes any regulation or any enactment in this Act, but if the work is one in respect of which plans were deposited and the plans were passed by the local authority, or notice of their rejection was not given within the prescribed period after the deposit thereof, and if the work has been executed in accordance with the plans, the court on granting an injunction shall have power to order the local authority to pay to the owner of the work such compensation as the court thinks just, but before making any such order the court shall in accordance with rules of court cause the local authority, if not a party to the proceedings, to be joined as a party thereto.

1961 c. 64.

The Public Health Act 1961

4. (1)

(2) Any provision contained in building regulations may be made so as to apply generally, or in an area specified in the regulations, and the regulations may make different provision for different areas and generally different provision for different circumstances or cases. SCH. 6

(3) It shall be the function of every local authority to enforce building regulations in their district.

(4) Local authorities shall, in relation to building regulations, have all such functions under sections 64 and 65 of the Public Health Act 1936 (which confer power to pass plans, and to enforce building byelaws) as they have in relation to building byelaws. 1936 c. 49.

(5) Building regulations may include such supplemental and incidental provisions as appear to the Secretary of State to be expedient.

(6) If a person contravenes or fails to comply with any provision contained in building regulations, other than a provision designated in the regulations as one to which this subsection does not apply, he shall be liable to a fine not exceeding £400 and to a further fine not exceeding £50 for each day on which the default continues after he is convicted.

(7) The power of making building regulations shall be exercisable by statutory instrument which shall be subject to annulment in pursuance of a resolution of either House of Parliament.

6.—(1) Subject to the provisions of this section, if the Secretary of State, on an application made in accordance with the provisions of this Act, considers that the operation of any requirement in building regulations would be unreasonable in relation to the particular case to which the application relates, he may after consultation with the local authority, give a direction dispensing with or relaxing that requirement either unconditionally or subject to compliance with any conditions specified in the direction, being conditions with respect to matters directly connected with the dispensation or relaxation.

(2) If building regulations so provide as regards any requirement contained in the regulations, the power to dispense with or relax that requirement under subsection (1) of this section shall be exercisable by the local authority (instead of by the Secretary of State after consultation with the local authority):

Provided than any building regulations made by virtue of this subsection may except applications of any description.

(3) Building regulations may provide as regards any requirement contained in the regulations that the foregoing subsections of this section shall not apply.

(4) An application under this section shall be in such form . . . . . . and shall contain such particulars as may be . . . . . . prescribed.

(5) The application shall be made to the local authority and, except where the power of giving the direction is exercisable by the local authority, the local authority shall at once transmit the application to the Secretary of State and give notice to the applicant that it has been so transmitted.

SCH. 6

(6) An application by a local authority in connection with a building or proposed building in the area of that authority shall be made to the Secretary of State except where the power of giving the direction is exercisable by that authority.

(7) The provisions of Part I of the First Schedule to this Act shall have effect as regards any application made under this section for a direction which will affect the application of building regulations to work which has been carried out before the making of the application.

(7A) If, on an application to the Secretary of State for a direction under this section, the Secretary of State considers that any requirement of building regulations to which the application relates is not applicable or is not or would not be contravened in the case of the work or proposed work to which the application relates he may so determine and may give any directions that he considers necessary in the circumstances.

(7B) A person who contravenes any condition specified in a direction given under this section or permits any such condition to be contravened shall be liable to a fine not exceeding £400 and to a further fine not exceeding £50 for each day on which the offence continues after he is convicted.

. . . . . .

7.—(1) If a local authority refuse an application to dispense with or relax any requirement in building regulations which they have power to dispense with or relax, or grant such an application subject to conditions, the applicant may in the prescribed manner appeal to the Secretary of State within the prescribed period from the date on which the local authority notify the applicant of their decision on the application.

(2) If within the prescribed period the local authority do not notify the applicant of their decision on the application, subsection (1) of this section shall apply in relation to the application as if the local authority had refused the application and notified the applicant of their decision at the end of the said period.

. . . .

(7) Section 6(7A) of this Act shall apply in relation to an appeal to the Secretary of State under this section as it applies in relation to an application to him for a direction under section 6.

Section 75.

SCHEDULE 7

1959 c. 24.

AMENDMENTS OF BUILDING (SCOTLAND) ACT 1959

1. In section 3 (building standards regulations)—

(*a*) in subsection (2), after the words " health, safety " there shall be inserted the word " welfare ", and at the end there shall be added the words " and for furthering the conservation of fuel and power " ;

(*b*) in subsection (3), there shall be added the words— Sch. 7

"(*d*) be framed to any extent by reference to a document published by or on behalf of the Secretary of State or any other person." ;

(*c*) at the end of the section there shall be added the following subsection—

"(7) The Secretary of State may by order made by statutory instrument repeal or modify any enactment to which this subsection applies if it appears to him that the enactment is inconsistent with, or is unnecessary or requires alteration in consequence of, any provision contained in the building standards regulations.

This subsection applies to any enactment contained in any Act passed before or in the same Session as the Health and Safety at Work etc. Act 1974 other than an enactment contained in the Building (Scotland) Act 1959."

2. In section 4 (relaxation of building standards regulations)—

(*a*) for subsection (5) there shall be substituted the following subsections—

"(5) A direction under subsection (1)(*b*) above—

(*a*) shall, if it so provides, cease to have effect at the end of such period as may be specified in the direction ;

(*b*) may be varied or revoked by a subsequent direction of the Secretary of State.

(5A) If at any time a direction under subsection (1)(*b*) above ceases to have effect by virtue of subsection (5)(*a*) above or is varied or revoked under subsection (5)(*b*) above, that fact shall not affect the continued operation of the direction (with any conditions specified therein) in any case in which before that time an application for a warrant in connection with the construction or change of use of a building, part or all of which is of the class to which the direction relates, was, in accordance with regulations made under section 2 of this Act, lodged with a building authority." ;

(*b*) in subsections (6) and (7), after the words "subsection (1)(*b*)" there shall be inserted the words "or (5)(*b*)" ;

(*c*) after subsection (7) there shall be inserted the following subsection :—

"(7A) A person making an application under subsection (1)(*b*) above shall pay to the Secretary of State such fee as may be prescribed ; and regulations made by virtue of this subsection may prescribe different fees for different cases :

Provided that the Secretary of State may in any particular case remit the whole or part of any fee payable by virtue of this subsection.".

SCH. 7

3. After section 4A, there shall be inserted the following section—

"Power of Secretary of State to approve types of building, etc.

4B.—(1) The following provisions of this section shall have effect with a view to enabling the Secretary of State, either on an application made to him in that behalf or of his own accord, to approve any particular type of building as conforming, either generally or in any class of case, to particular provisions of the building standards regulations.

(2) An application for the approval under this section of a type of building shall be made in the prescribed manner.

(3) Where under subsection (1) above the Secretary of State approves a type of building as conforming to particular provisions of the building standards regulations either generally or in any class of case, he may issue a certificate to that effect specifying—

(*a*) the type of building to which the certificate relates ;

(*b*) the provisions of the building standards regulations to which the certificate relates ; and

(*c*) where applicable, the class or classes of case to which the certificate applies.

(4) A certificate under this section shall, if it so provides, cease to have effect at the end of such period as may be specified in the certificate.

(5) If, while a certificate under this section is in force, it is found, in any particular case involving a building of the type to which the certificate relates, that the building in question is of that type and the case is one to which the certificate applies, that building shall in that particular case be deemed to conform to the provisions of the building standards regulations to which the certificate relates.

(6) The Secretary of State may from time to time vary a certificate under this section either on an application made to him in that behalf or of his own accord ; but in the case of a certificate issued on an application made by a person under subsection (1) above, the Secretary of State, except where he varies it on the application of that person, shall before varying it give that person reasonable notice that he proposes to do so.

(7) A person making an application under subsection (1) or (6) above shall pay to the Secretary of State such fee as may be prescribed ; and regulations made by virtue of this subsection may prescribe different fees for different cases :

Provided that the Secretary of State may in any particular case remit the whole or part of any fee payable by virtue of this subsection.

(8) The Secretary of State may at any time revoke a certificate issued under this section, but before doing so shall give the person, if any, on whose application the certificate was issued reasonable notice that he proposes to do so.

(9) Where the Secretary of State issues a certificate under this section or varies or revokes a certificate so issued, he shall publish notice of that fact in such manner as he thinks fit.

(10) If at any time a certificate under this section ceases to have effect by virtue of subsection (4) above or is varied or revoked under the preceding provisions of this section, that fact shall not affect the continued operation of subsection (5) above by virtue of that certificate in any case in which before that time an application for a warrant in connection with the construction of a type of building to which the certificate relates was, in accordance with regulations made under section 2 of this Act, lodged with a buildings authority.

(11) For the purposes of subsection (3) above or any variation of a certificate under subsection (6) above, a class of case may be framed in any way that the Secretary of State thinks fit."

4. In section 6 (application of building standards regulations and building operations regulations to construction or demolition, and to change of use, of buildings)—

(*a*) after subsection (3) there shall be inserted the following subsection—

"(3A) Notwithstanding that a buildings authority are not satisfied that the information submitted to them with an application for a warrant for the construction of a building is sufficient in respect of such stage in the construction as may be prescribed to show that the building when constructed will not fail to conform to the building standards regulations, they may grant a warrant for the construction of the building but subject to the condition that work on such prescribed stage shall not be proceeded with until such further information relating to that stage as they may require is submitted to them and until they have made an amendment to the terms of the warrant authorising such work to proceed:

Provided that they shall, subject to subsection (8) of this section, make such an amendment on application being made therefor in the prescribed manner only if they are satisfied that nothing in the information submitted to them in respect of the prescribed stage shows that that stage when constructed will fail to conform to the building standards regulations.";

(*b*) in subsection (10), after the words "any such" there shall be inserted the words "prescribed stage as is mentioned in subsection (3A) of this section and any such".

SCH. 7

5. In section 9 (certificates of completion)—

(*a*) in subsection (2), for the words " but only if, they are satisfied that " there shall be substituted the words " , so far as they are able to ascertain after taking all reasonable steps in that behalf," ;

(*b*) in subsection (3), for the words, " be satisfied as mentioned in the last foregoing subsection " there shall be substituted the words " grant a certificate of completion " ;

(*c*) after subsection (3) there shall be inserted the following subsection—

" (3A) In respect of so much of a building as consists of such an installation as may be prescribed, not being an electrical installation, a buildings authority shall not grant a certificate of completion unless there is produced to them a certificate granted by a person of such class as may be prescribed certifying that the installation complies with such of the said conditions as relate to it:

Provided that this subsection shall not apply in a case where it is shown to the satisfaction of the buildings authority that for some reasonable cause such a certificate cannot be produced." ;

(*d*) in subsection (4) for the words " the last foregoing subsection " there shall be substituted the words " subsection (3) or (3A) above ".

6. In section 11(1)(*b*) (power of local authorities to require buildings to conform to building standards regulations), after the words " health, safety " there shall be inserted the word " welfare ", and after the word " generally " there shall be inserted the words " and for furthering the conservation of fuel and power ".

7. In section 19 (penalties), for the words " ten pounds " and " one hundred pounds ", wherever they occur, there shall be substituted respectively the words " £50 " and " £400 ".

8. After section 19 there shall be inserted the following section—

" Civil liability.

19A.—(1) Subject to the provisions of this section, a breach to which this section applies shall, so far as it causes damage, be actionable except in so far as may be otherwise prescribed ; and in any action brought by virtue of this subsection such defence as may be prescribed shall be available.

(2) This section applies to the following breaches—

(a) failure to comply with the terms or conditions of a warrant for the construction, demolition or change of use of a building or with any order under this Act relating to the construction of a building ;

(*b*) contravention of any provision of the building operations regulations ;

(*c*) constructing a building without a warrant otherwise than in accordance with the building standards regulations ;

(*d*) changing the use of a building without a warrant where after the change of use the building does not conform to so much of the building standards regulations as become applicable, or apply more onerously, to the building by reason of the change of use.

(3) Subsection (1) above and any defence provided for in regulations made by virtue thereof shall not apply in the case of a breach to which this section applies in connection with a building erected before the date on which that subsection comes into force unless the breach arises in relation to the change of use, extension, alteration, demolition, repair, maintenance or fitting of such a building.

(4) Nothing in this section shall be construed as affecting the extent (if any) to which a breach to which this section applies is actionable in a case to which subsection (1) above does not apply, or as prejudicing any right of action which exists apart from the provisions of this section.

(5) In this section " damage " includes the death of, or injury to, any person (including any disease and any impairment of a person's physical or mental condition)."

9. In section 26 (Crown rights)—

(*a*) in subsection (1) after the words " Crown and " there shall be inserted the words " subject to the provisions of this section " ;

(*b*) after subsection (2) there shall be inserted the following subsections—

" (2A) The building standards regulations shall, except in so far as they otherwise prescribe, apply to a Crown building as they would apply if the building were not a Crown building.

(2B) A Crown building to which the building standards regulations apply shall be constructed in accordance with those regulations.

(2C) Any extension to or alteration of a Crown building to which the building standards regulations apply or would apply on the extension or alteration of the building shall not cause the building as extended or altered, as a direct result of the extension or, as the case may be, the alteration—

(*a*) if it conformed to the building standards regulations immediately before the date of commencement of the operations, to fail to conform to them ; or

SCH. 7

(*b*) if it failed to conform to the building standards regulations immediately before that date, to fail to conform to them to a greater degree than that to which it failed to conform immediately before that date ;

and any change of use of a Crown building shall not cause the building after the change of use to fail to conform to so much of the building standards regulations as will become applicable, or will apply more onerously, to the building by reason of the change of use.

(2D) Section 19A of this Act shall apply to a Crown building as it applies to a building other than a Crown building, but as if for subsection (2) there were substituted the following subsection :—

" (2) A breach to which this section applies is a failure to comply with subsection (2B) or (2C) of section 26 of this Act or a contravention of any provision of the building operations regulations ".

(2E) Without prejudice to any case to which proviso (*a*) to subsection (1) above is applicable, the Secretary of State shall have the like powers of dispensing with or relaxing the provisions of the building standards regulations in relation to a Crown building as he has under section 4(1) of this Act in relation to a building other than a Crown building ; and subsections (3), (4), (5), (5A) and (9) of the said section 4 shall apply for the purposes of this section as if—

(*a*) in subsection (4), the words " or, as the case may be, the buildings authority " were omitted ;

(*b*) in subsection (5A), for the words from " an application " to the end there were substituted the words " the construction or change of use of a building, part or all of which is of the class to which the direction relates, was begun " ;

(*c*) in subsection (9), the words " or section 4A(3) of this Act " were omitted.

(2F) Without prejudice to any case to which the said proviso is applicable, in the application of section 4B of this Act to a Crown building, subsection (10) shall have effect as if for the words from " an application " to the end there were substituted the words " the construction of a building, part or all of which is of the type to which the certificate relates, was begun "."

Section 78.

SCHEDULE 8

1961 c. 34.
1963 c. 41.

TRANSITIONAL PROVISIONS WITH RESPECT TO FIRE CERTIFICATES UNDER FACTORIES ACT 1961 OR OFFICES, SHOPS AND RAILWAY PREMISES ACT 1963

1. In this Schedule—

1971 c. 40.

" the 1971 Act " means the Fire Precautions Act 1971 ;

" 1971 Act certificate " means a fire certificate within the meaning of the 1971 Act ;

SCH. 8

"Factories Act certificate" means a certificate under section 40 of the Factories Act 1961 (means of escape in case of fire-certification by fire authority); 1961 c. 34.

"Offices Act certificate" means a fire certificate under section 29 of the Offices, Shops and Railway Premises Act 1963. 1963 c. 41.

2.—(1) Where by virtue of an order under section 1 of the 1971 Act a 1971 Act certificate becomes required in respect of any premises at a time when there is in force in respect of those premises a Factories Act certificate or an Offices Act certificate ("the existing certificate"), the following provisions of this paragraph shall apply.

(2) The existing certificate shall continue in force (irrespective of whether the section under which it was issued remains in force) and—

(a) shall as from the said time be deemed to be a 1971 Act certificate validly issued with respect to the premises with respect to which it was issued and to cover the use or uses to which those premises were being put at that time; and

(b) may (in particular) be amended, replaced or revoked in accordance with the 1971 Act accordingly.

(3) Without prejudice to sub-paragraph (2)(b) above, the existing certificate, as it has effect by virtue of sub-paragraph (2) above, shall as from the said time be treated as imposing in relation to the premises the like requirements as were previously imposed in relation thereto by the following provisions, that is to say—

(a) if the existing certificate is a Factories Act certificate, the following provisions of the Factories Act 1961, namely sections 41(1), 48 (except subsections (5), (8) and (9)), 49(1), 51(1) and 52(1) and (4) and, so far as it relates to a proposed increase in the number of persons employed in any premises, section 41(3);

(b) if the existing certificate is an Offices Act certificate the following provisions of the Offices, Shops and Railway Premises Act 1963, namely sections 30(1), 33, 34(1) and (2), 36(1) and 38(1) and, so far as it relates to a proposed increase in the number of persons employed to work in any premises at any one time, section 30(3).

3. Any application for a Factories Act certificate or an Offices Act certificate with respect to any premises which is pending at the time when by virtue of an order under section 1 of the 1971 Act a 1971 Act certificate becomes required in respect of those premises shall be deemed to be an application for a 1971 Act certificate in respect of them duly made in accordance with the 1971 Act and may be proceeded with accordingly; but (without prejudice to section 5(2) of the 1971 Act) the fire authority may, as a condition of proceeding with such an application, require the applicant to specify any matter or give them any information which would ordinarily have been required by section 5(1) of that Act.

Section 83.

SCHEDULE 9

Minor and Consequential Amendments

1926 c. 59.

The Coroners (Amendment) Act 1926

1. In section 13(2)(*c*) of the Coroners (Amendment) Act 1926 (by virtue of which an inquest must be held with a jury in cases of death from certain causes of which notice is required to be given to any inspector or other officer of a government department), after the words " of a government department " there shall be inserted the words " or to an inspector appointed under section 19 of the Health and Safety at Work etc. Act 1974,".

1957 c. 20.

The House of Commons Disqualification Act 1957

2. In Part II of Schedule 1 to the House of Commons Disqualification Act 1957 (which specifies bodies of which all members are disqualified under that Act), as it applies to the House of Commons of the Parliament of the United Kingdom, there shall be inserted at the appropriate place in alphabetical order the words " The Health and Safety Commission ".

1967 c. 13.

The Parliamentary Commissioner Act 1967

3. In Schedule 2 to the Parliamentary Commissioner Act 1967 (which lists the authorities subject to investigation under that Act) there shall be inserted in the appropriate places in alphabetical order the words " Health and Safety Commission " and " Health and Safety Executive ".

Section 83.

SCHEDULE 10

Repeals

Chapter	Short Title	Extent of repeal
26 Geo. 5 & 1 Edw. 8. c. 49.	The Public Health Act 1936.	Section 53. Section 64(4) and (5). In section 67, the words from " and the Secretary of State's decision " to the end of the section. Section 71. In section 343(1), the definition of " building regulations ".
7 & 8 Geo. 6. c. 31.	The Education Act 1944.	Section 63(1).
10 & 11 Geo. 6. c. 51.	The Town and Country Planning Act 1947.	In Schedule 8, the amendment of section 53 of the Public Health Act 1936.
2 & 3 Eliz. 2. c. 32.	The Atomic Energy Authority Act 1954.	Section 5(5).
4 & 5 Eliz. 2. c. 52.	The Clean Air Act 1956.	Section 24.

Chapter	Short Title	Extent of Repeal
9 & 10 Eliz. 2. c. 64.	The Public Health Act 1961.	In section 4, subsection (1) and, in subsection (4), the words from " and building " to the end of the subsection. In section 6, in subsection (4), the words " as may be prescribed by building regulations " and the word " so ", and subsection (8). Section 7(3) to (6). Section 10(1) and (2). In Schedule 1, in Part III, the amendments of sections 53, 61, 62 and 71 of the Public Health Act 1936 and, in the amendments of the Clean Air Act 1956, the amendment of section 24 and the word " twenty-four " in the last paragraph.
1965 c. 16.	The Airports Authority Act 1965.	In section 19(3), the words from " and section 71 " to " regulations) " and the words " and the proviso to the said section 71 ".
1971 c. 40.	The Fire Precautions Act 1971.	In section 2, paragraphs (*a*) to (*c*). Section 11. In section 17(1)(i), the word " and " where last occurring. In section 43(1), the definition of " building regulations ".
1971 c. 75.	The Civil Aviation Act 1971.	In Schedule 5, in paragraph 2(1), the words from " and section 71 " to " regulations)" and the words " and the proviso to the said section 71 ".
1972 c. 28.	The Employment Medical Advisory Service Act 1972.	Sections 1 and 6. Schedule 1.
1972 c. 58.	The National Health Service (Scotland) Act 1972.	In Schedule 6, paragraph 157.
1972 c. 70.	The Local Government Act 1972.	In Schedule 14, paragraph 43.
1973 c. 32.	The National Health Service Reorganisation Act 1973.	In Schedule 4, paragraph 137.
1973 c. 50.	The Employment and Training Act 1973.	In Schedule 3, paragraph 14.
1973 c. 64.	The Maplin Development Act 1973.	In Schedule 2, in paragraph 2(1), the words from " and section 71 " to " regulations) ".

PRODUCED IN ENGLAND BY COMMERCIAL COLOUR PRESS LONDON
FOR BERNARD M THIMONT
Controller of Her Majesty's Stationery Office and Queen's Printer of Acts of Parliament

Dd.593996 K240 10/80

Appendix I
Part B

tion 116.
4 c. 37.

SCHEDULE 15

AMENDMENTS OF THE HEALTH AND SAFETY AT WORK ETC. ACT 1974

1. In section 1(2) omit the words " and agricultural health and safety regulations ".

2. In section 2, omit subsection (5) and in subsection (7) for the words " subsections (4) and (5) " substitute the words " subsection (4) ".

3. After section 10(7) insert the following subsection:—

" (8) For the purposes of any civil proceedings arising out of those functions, the Crown Proceedings Act 1947 and the Crown Suits (Scotland) Act 1857 shall apply to the Commission and the Executive as if they were government departments within the meaning of the said Act of 1947 or, as the case may be, public departments within the meaning of the said Act of 1857.".

4. In section 11, in subsection (1) omit the words " except as regards matters relating exclusively to agricultural operations ", and in subsection (2) omit the words " except as aforesaid ".

5. In section 14(2), omit the words from " but shall not do so " to " agricultural operations ".

6. In section 15, for subsection (1) substitute—

" (1) Subject to the provisions of section 50, the Secretary of State, the Minister of Agriculture, Fisheries and Food or the Secretary of State and that Minister acting jointly shall have power to make regulations under this section for any of the general purposes of this Part (and regulations so made are in this Part referred to as " health and safety regulations ").".

7. In section 16(1), omit the words " and except as regards matters relating exclusively to agricultural operations ".

8. In section 18, in subsection (3) omit the words " or agricultural health and safety regulations ", and in subsection (5) omit the words " the appropriate Agriculture Minister ".

9. In section 28, after subsection (8) insert the following subsection—

"(9) Notwithstanding anything in subsection (7) above, a person who has obtained such information as is referred to in that subsection may furnish to a person who appears to him to be likely to be a party to any civil proceedings arising out of any accident, occurrence, situation or other matter, a written statement of relevant facts observed by him in the course of exercising any of the powers referred to in that subsection.".

10. Sections 29, 30, 31 and 32 are hereby repealed.

11. In section 33, in subsection (1)(*c*) omit the words "or agricultural health and safety regulations", and in subsection (4)(*a*) omit the words "or the appropriate Agriculture Minister".

12. In section 43, in subsection (3) omit the words "the Minister of Agriculture, Fisheries and Food" and for subsections (6) and (7) substitute—

"(6) The power to make regulations under this section shall be exercisable by the Secretary of State, the Minister of Agriculture, Fisheries and Food or the Secretary of State and that Minister acting jointly.".

13. In section 44, in subsection (1) omit the words "agricultural licences and", and in subsection (7)(*a*) for the words "an agricultural licence or nuclear site licence" substitute the words "a nuclear site licence".

14. In section 47, in subsection (2) omit the words "or agricultural health and safety regulations", in subsection (3) omit the words "or, as the case may be, agricultural health and safety regulations" and in subsection (5) omit the words "or, as the case may be, agricultural health and safety regulations".

15.—(1) In section 49, in subsection (1) for the words "The appropriate Minister may by regulations amend" substitute the words "Regulations made under this subsection may amend", in subsection (2) for the words "appropriate Minister" substitute the words "authority making the regulations", in subsection (3) omit the words "by the appropriate Minister" and for the words "if the appropriate Minister" substitute the words "if the authority making the regulations".

(2) For subsection (4) of that section substitute—

"(4) The power to make regulations under this section shall be exercisable by the Secretary of State, the Minister of Agriculture, Fisheries and Food or the Secretary of State and that Minister acting jointly.".

16.—(1) In section 50, for subsection (1) substitute—

"(1) Where any power to make regulations under any of the relevant statutory provisions is exercisable by the Secretary of State, the Minister of Agriculture, Fisheries and Food or both of them acting jointly that power may be exercised either so as to give effect (with or without modifications) to proposals submitted by the Commission under section 11(2)(*d*) or independently of any such proposals; but the authority who is to exercise the power shall not exercise it independently of proposals from the Commission unless he has consulted the Commission and such other bodies as appear to him to be appropriate.".

(2) In subsection (2) of that section, for the words from "Secretary of State" to "preceding subsection" substitute "authority who is to exercise any such power as is mentioned in subsection (1) above proposes to exercise that power".

(3) In subsection (3), for the words "to the Secretary of State" substitute the words "under section 11(2)(*d*)".

(4) Subsections (4) and (5) are hereby repealed.

17. In section 52, for subsections (3) and (4) substitute—

"(3) The power to make regulations under subsection (2) above shall be exercisable by the Secretary of State, the Minister of Agriculture, Fisheries and Food or the Secretary of State and that Minister acting jointly.".

18.—(1) In section 53, in subsection (1) omit the definitions of "agriculture", "the Agriculture Ministers", "agricultural health and safety regulations", "agricultural licence", "agricultural operation", "the appropriate Agriculture Minister", "forestry", "livestock" and "the relevant agricultural purposes" and in the definition of "the relevant statutory provisions" omit the words "and agricultural health and safety regulations".

(2) Subsections (2) to (6) of that section are hereby repealed.

19. In section 80, for subsections (4) to (6) substitute—

"(4) The power to make regulations under subsection (1) above shall be exercisable by the Secretary of State, the Minister of Agriculture, Fisheries and Food or the Secretary of State and that Minister acting jointly; but the authority who is to exercise the power shall, before exercising it, consult such bodies as appear to him to be appropriate.

(5) In this section 'the relevant statutory provisions' has the same meaning as in Part I.".

20. In section 84(1)(*a*), omit the words "or 30".

21. Schedule 4 is hereby repealed.

Appendix I
Part C

The Safety Representatives and Safety Committees Regulations 1977

Made - - - -	16*th March* 1977
Laid before Parliament	28*th March* 1977
Coming into Operation	1*st October* 1978

Whereas the Health and Safety Commission has submitted to the Secretary of State, under section 11(2)(*d*) of the Health and Safety at Work etc. Act 1974(**a**) ("the 1974 Act") as amended by paragraph 4 of Schedule 15 to the Employment Protection Act 1975(**b**) ("the 1975 Act"), proposals for the making of Regulations after the carrying out by the said Commission of consultations in accordance with section 50(3) of the 1974 Act as amended by paragraph 16(3) of Schedule 15 to the 1975 Act;

And whereas the Secretary of State has made modifications to the said proposals under section 50(1) of the 1974 Act and has consulted the said Commission thereon in accordance with section 50(2) of the 1974 Act, both of which provisions have been amended by paragraph 16 of Schedule 15 to the 1975 Act;

And whereas under section 80(1) of the 1974 Act it appears to the Secretary of State that the modification of paragraph 16 of Schedule 1 to the Trade Union and Labour Relations Act 1974(**c**) made in Regulation 11(5) below is expedient in connection with the provision made by Regulation 4(2) below and, in accordance with section 80(4) of the 1974 Act as substituted by paragraph 19 of Schedule 15 to the 1975 Act, he has consulted such bodies as appeared to him to be appropriate on the proposed modification;

Now therefore, the Secretary of State, in exercise of the powers conferred on him by section 2(4) and (7), 15(1), (3)(*b*) and (5)(*b*), 80(1) and (4) and 82(3)(*a*) of the 1974 Act as amended by paragraphs 2, 6 and 19 of Schedule 15 to the 1975 Act and of all other powers enabling him in that behalf and so as to give effect to the said proposals of the said Commission (with the said modifications) and so as to modify the said provision in the Trade Union and Labour Relations Act 1974, hereby makes the following Regulations: —

Citation and commencement

1. These Regulations may be cited as the Safety Representatives and Safety Committees Regulations 1977 and shall come into operation on 1st October 1978.

(**a**) 1974 c. 37. (**b**) 1975 c. 71. (**c**) 1974 c. 52.

Interpretation

2.—(1) In these Regulations, unless the context otherwise requires—

"the 1974 Act" means the Health and Safety at Work etc. Act 1974 as amended by the 1975 Act;

"the 1975 Act" means the Employment Protection Act 1975;

"employee" has the meaning assigned by section 53(1) of the 1974 Act and "employer" shall be construed accordingly;

"recognised trade union" means an independent trade union as defined in section 30(1) of the Trade Union and Labour Relations Act 1974 which the employer concerned recognises for the purpose of negotiations relating to or connected with one or more of the matters specified in section 29(1) of that Act in relation to persons employed by him or as to which the Advisory, Conciliation and Arbitration Service has made a recommendation for recognition under the 1975 Act which is operative within the meaning of section 15 of that Act;

"safety representative" means a person appointed under Regulation 3(1) of these Regulations to be a safety representative;

"welfare at work" means those aspects of welfare at work which are the subject of health and safety regulations or of any of the existing statutory provisions within the meaning of section 53(1) of the 1974 Act;

"workplace" in relation to a safety representative means any place or places where the group or groups of employees he is appointed to represent are likely to work or which they are likely to frequent in the course of their employment or incidentally to it.

(2) The Interpretation Act 1889(**a**) shall apply to the interpretation of these Regulations as it applies to the interpretation of an Act of Parliament.

(3) These Regulations shall not be construed as giving any person a right to inspect any place, article, substance or document which is the subject of restrictions on the grounds of national security unless he satisfies any test or requirement imposed on those grounds by or on behalf of the Crown.

Appointment of safety representatives

3.—(1) For the purposes of section 2(4) of the 1974 Act, a recognised trade union may appoint safety representatives from amongst the employees in all cases where one or more employees are employed by an employer by whom it is recognised, except in the case of employees employed in a mine within the meaning of section 180 of the Mines and Quarries Act 1954(**b**) which is a coal mine.

(2) Where the employer has been notified in writing by or on behalf of a trade union of the names of the persons appointed as safety representatives under this Regulation and the group or groups of employees they represent, each such safety representative shall have the functions set out in Regulation 4 below.

(**a**) 1889 c. 63. (**b**) 1954 c. 70.

(3) A person shall cease to be a safety representative for the purposes of these Regulations when—

(*a*) the trade union which appointed him notifies the employer in writing that his appointment has been terminated; or

(*b*) he ceases to be employed at the workplace but if he was appointed to represent employees at more than one workplace he shall not cease by virtue of this sub-paragraph to be a safety representative so long as he continues to be employed at any one of them; or

(*c*) he resigns.

(4) A person appointed under paragraph (1) above as a safety representative shall so far as is reasonably practicable either have been employed by his employer throughout the preceding two years or have had at least two years experience in similar employment.

Functions of safety representatives

4.—(1) In addition to his function under section 2(4) of the 1974 Act to represent the employees in consultations with the employer under section 2(6) of the 1974 Act (which requires every employer to consult safety representatives with a view to the making and maintenance of arrangements which will enable him and his employees to cooperate effectively in promoting and developing measures to ensure the health and safety at work of the employees and in checking the effectiveness of such measures), each safety representative shall have the following functions:—

(*a*) to investigate potential hazards and dangerous occurrences at the workplace (whether or not they are drawn to his attention by the employees he represents) and to examine the causes of accidents at the workplace;

(*b*) to investigate complaints by any employee he represents relating to that employee's health, safety or welfare at work;

(*c*) to make representations to the employer on matters arising out of sub-paragraphs (*a*) and (*b*) above;

(*d*) to make representations to the employer on general matters affecting the health, safety or welfare at work of the employees at the workplace;

(*e*) to carry out inspections in accordance with Regulations 5, 6 and 7 below;

(*f*) to represent the employees he was appointed to represent in consultations at the workplace with inspectors of the Health and Safety Executive and of any other enforcing authority;

(*g*) to receive information from inspectors in accordance with section 28(8) of the 1974 Act; and

(*h*) to attend meetings of safety committees where he attends in his capacity as a safety representative in connection with any of the above functions;

but, without prejudice to sections 7 and 8 of the 1974 Act, no function given to a safety representative by this paragraph shall be construed as imposing any duty on him.

(2) An employer shall permit a safety representative to take such time off with pay during the employee's working hours as shall be necessary for the purposes of—

(*a*) performing his functions under section 2(4) of the 1974 Act and paragraph (1)(*a*) to (*h*) above;

(*b*) undergoing such training in aspects of those functions as may be reasonable in all the circumstances having regard to any relevant provisions of a code of practice relating to time off for training approved for the time being by the Health and Safety Commission under section 16 of the 1974 Act.

In this paragraph "with pay" means with pay in accordance with the Schedule to these Regulations.

Inspections of the workplace

5.—(1) Safety representatives shall be entitled to inspect the workplace or a part of it if they have given the employer or his representative reasonable notice in writing of their intention to do so and have not inspected it, or that part of it, as the case may be, in the previous three months; and may carry out more frequent inspections by agreement with the employer.

(2) Where there has been a substantial change in the conditions of work (whether because of the introduction of new machinery or otherwise) or new information has been published by the Health and Safety Commission or the Health and Safety Executive relevant to the hazards of the workplace since the last inspection under this Regulation, the safety representatives after consultation with the employer shall be entitled to carry out a further inspection of the part of the workplace concerned notwithstanding that three months have not elapsed since the last inspection.

(3) The employer shall provide such facilities and assistance as the safety representatives may reasonably require (including facilities for independent investigation by them and private discussion with the employees) for the purpose of carrying out an inspection under this Regulation, but nothing in this paragraph shall preclude the employer or his representative from being present in the workplace during the inspection.

(4) An inspection carried out under section 123 of the Mines and Quarries Act 1954 shall count as an inspection under this Regulation.

Inspections following notifiable accidents, occurrences and diseases

6.—(1) Where there has been a notifiable accident or dangerous occurrence in a workplace or a notifiable disease has been contracted there and—

(*a*) it is safe for an inspection to be carried out; and

(*b*) the interests of employees in the group or groups which safety representatives are appointed to represent might be involved.

those safety representatives may carry out an inspection of the part of the workplace concerned and so far as is necessary for the purpose of determining the cause they may inspect any other part of the workplace; where it is reasonably practicable to do so they shall notify the employer or his representative of their intention to carry out the inspection.

(2) The employer shall provide such facilities and assistance as the safety representatives may reasonably require (including facilities for independent investigation by them and private discussion with the employees) for the purpose of carrying out an inspection under this Regulation; but nothing in this paragraph shall preclude the employer or his representative from being present in the workplace during the inspection.

(3) In this Regulation "notifiable accident or dangerous occurrence" and "notifiable disease" mean any accident, dangerous occurrence or disease, as the case may be, notice of which is required to be given by virtue of any of the relevant statutory provisions within the meaning of section 53(1) of the 1974 Act.

Inspection of documents and provision of information

7.—(1) Safety representatives shall for the performance of their functions under section 2(4) of the 1974 Act and under these Regulations, if they have given the employer reasonable notice, be entitled to inspect and take copies of any document relevant to the workplace or to the employees the safety representatives represent which the employer is required to keep by virtue of any relevant statutory provision within the meaning of section 53(1) of the 1974 Act except a document consisting of or relating to any health record of an identifiable individual.

(2) An employer shall make available to safety representatives the information, within the employer's knowledge, necessary to enable them to fulfil their functions except—

(*a*) any information the disclosure of which would be against the interests of national security; or

(*b*) any information which he could not disclose without contravening a prohibition imposed by or under an enactment; or

(c) any information relating specifically to an individual, unless he has consented to its being disclosed; or

(*d*) any information the disclosure of which would, for reasons other than its effect on health, safety or welfare at work, cause substantial injury to the employer's undertaking or, where the information was supplied to him by some other person, to the undertaking of that other person; or

(*e*) any information obtained by the employer for the purpose of bringing, prosecuting or defending any legal proceedings.

(3) Paragraph (2) above does not require an employer to produce or allow inspection of any document or part of a document which is not related to health, safety or welfare.

Cases where safety representatives need not be employees

8.—(1) In the cases mentioned in paragraph (2) below safety representatives appointed under Regulation 3(1) of these Regulations need not be employees of the employer concerned; and section 2(4) of the 1974 Act shall be modified accordingly.

(2) The said cases are those in which the employees in the group or groups the safety representatives are appointed to represent are members of the British Actors' Equity Association or of the Musicians' Union.

(3) Regulations 3(3)(*b*) and (4) and 4(2) of these Regulations shall not apply to safety representatives appointed by virtue of this Regulation and in the case of safety representatives to be so appointed Regulation 3(1) shall have effect as if the works "from amongst the employees" were omitted.

Safety committees

9.—(1) For the purposes of section 2(7) of the 1974 Act (which requires an employer in prescribed cases to establish a safety committee if requested to do so by safety representatives) the prescribed cases shall be any cases in which at least two safety representatives request the employer in writing to establish a safety committee.

(2) Where an employer is requested to establish a safety committee in a case prescribed in paragraph (1) above, he shall establish it in accordance with the following provisions—

(*a*) he shall consult with the safety representatives who made the request and with the representatives of recognised trade unions whose members work in any workplace in respect of which he proposes that the committee should function;

(*b*) the employer shall post a notice stating the composition of the committee and the workplace or workplaces to be covered by it in a place where it may be easily read by the employees;

(*c*) the committee shall be established not later than three months after the request for it.

Power of Health and Safety Commission to grant exemptions

10. The Health and Safety Commission may grant exemptions from any requirement imposed by these Regulations and any such exemption may be unconditional or subject to such conditions as the Commission may impose and may be with or without a limit of time.

Provisions as to industrial tribunals

11.—(1) A safety representative may, in accordance with the jurisdiction conferred on industrial tribunals by paragraph 16(2) of Schedule 1 to the Trade Union and Labour Relations Act 1974, present a complaint to an industrial tribunal that—

(*a*) the employer has failed to permit him to take time off in accordance with Regulation 4(2) of these Regulations; or

(*b*) the employer has failed to pay him in accordance with Regulation 4(2) of and the Schedule to these Regulations.

(2) An industrial tribunal shall not consider a complaint under paragraph (1) above unless it is presented within three months of the date when the failure occurred or within such further period as the tribunal considers reasonable in a case where it is satisfied that it was not reasonably practicable for the complaint to be presented within the period of three months.

(3) Where an industrial tribunal finds a complaint under paragraph (1)(*a*) above well-founded the tribunal shall make a declaration to that effect and may make an award of compensation to be paid by the employer to the employee which shall be of such amount as the tribunal considers just and equitable in all the circumstances having regard to the employer's default in failing to permit time off to be taken by the employee and to any loss sustained by the employee which is attributable to the matters complained of.

(4) Where on a complaint under paragraph (1)(*b*) above an industrial tribunal finds that the employer has failed to pay the employee the whole or part of the amount required to be paid under paragraph (1)(*b*), the tribunal shall order the employer to pay the employee the amount which it finds due to him.

(5) Paragraph 16 of Schedule 1 to the Trade Union and Labour Relations Act 1974 (jurisdiction of industrial tribunals) shall be modified by adding the following sub-paragraph:—

"(2) An industrial tribunal shall have jurisdiction to determine complaints relating to time off with pay for safety representatives appointed under regulations made under the Health and Safety at Work etc. Act 1974".

Albert Booth,

16th March 1977. Secretary of State for Employment.

Regulation 4(2)

SCHEDULE

Pay for time off allowed to safety representatives

1. Subject to paragraph 3 below, where a safety representative is permitted to take time off in accordance with Regulation 4(2) of these Regulations, his employer shall pay him—

(*a*) where the safety representative's remuneration for the work he would ordinarily have been doing during that time does not vary with the amount of work done, as if he had worked at that work for the whole of that time;

(*b*) where the safety representative's remuneration for that work varies with the amount of work done, an amount calculated by reference to the average hourly earnings for that work (ascertained in accordance with paragraph 2 below).

2. The average hourly earnings referred to in paragraph 1(*b*) above are the average hourly earnings of the safety representative concerned or, if no fair estimate can be made of those earnings, the average hourly earnings for work of that description of persons in comparable employment with the same employer or, if there are no such persons, a figure of average hourly earnings which is reasonable in the circumstances.

3. Any payment to a safety representative by an employer in respect of a period of time off–

(*a*) if it is a payment which discharges any liability which the employer may have under section 57 of the 1975 Act in respect of that period, shall also discharge his liability in respect of the same period under Regulation 4(2) of these Regulations;—

(*b*) if it is a payment under any contractual obligation, shall go towards discharging the employer's liability in respect of the same period under Regulation 4(2) of these Regulations;

(*c*) if it is a payment under Regulation 4(2) of these Regulations shall go towards discharging any liability of the employer to pay contractual remuneration in respect of the same period.

Appendix II

List of statutory materials referred to in the text

A. Statutes

The earlier statutory provisions are marked *

- * Explosives Act 1875
- * Boiler Explosions Acts 1882, 1890
- * Alkali, etc Works Regulation Act 1906
- * Revenue Act 1909
- * Anthrax Prevention Act 1919
- * Employment of Women, Young Persons and Children Act 1920
- * Celluloid and Cinematograph Film Act 1922
- * Explosives Act 1923
- * Public Health Act (Smoke Abatement) Act 1926
- * Petroleum (Consolidation) Act 1928
- * Hours of Employment (Conventions) Act 1936
- * Petroleum (Transfer of Licences) Act 1936
- * Hydrogen Cyanide (Fumigation) Act 1937
- Truck Acts 1831–1940
- * Ministry of Fuel and Power Act 1945
- * Coal Industry Nationalization Act 1946
- * Radioactive Substances Act 1948
- * Alkali, etc (Works) Regulation (Scotland) Act 1951
- * Fireworks Act 1951
- * Agriculture (Poisonous Substances) Act 1952
- * Emergency Laws (Miscellaneous Provisions) Act 1953
- * Bakery Industry (Hours of Work) Act 1954
- * Mines and Quarries Act 1954
- * Agriculture (Safety, Health and Welfare Provisions) Act 1956
- * Factories Act 1961
- * Public Health Act 1961
- * Pipe-lines Act 1962
- * Offices, Shops and Railway Premises Act 1963

* Nuclear Installations Act 1965
* Mines and Quarries (Tips) Act 1969
  Employers' Liability (Compulsory Insurance) Act 1969
  The Occupational Safety and Health Act 1970 (USA)
  Mineral Workings (Offshore Installations) Act 1971
* Mines (Management) Act 1971
* Employment Medical Advisory Service Act 1972
  European Communities Act 1972
  Road Traffic Act 1972
  Control of Pollution Act 1974
  Trade Union and Labour Relations Act 1974
  Employment Protection Act 1975
  Sex Discrimination Act 1975
  Trade Union and Labour Relations (Amendment) Act 1976
  Employment Protection (Consolidation) Act 1978
  Employment Act 1980

B. Regulations

The Agriculture (Avoidance of Accidents to Children) Regulations 1958 No. 366

The Quarries (Explosives) Order 1959 No. 2259

The Ionising Radiations (Sealed Sources) Regulations 1961 No. 1470

The Motor Vehicles (Construction and Use) Regulations 1963 No. 1646

The Industrial Tribunals (Industrial Relations) Regulations 1972 No. 38

The Industrial Tribunals (Improvement and Prohibition Notices Appeals) Regulations 1974 No. 1925

The Woodworking Machines Regulations 1974 No. 903

The Employers' Health and Safety Policy Statements (Exceptions) Regulations 1975 No. 1584

The Safety Representatives and Safety Committees Regulations 1977 No. 500

The Health and Safety at Work etc. Act 1974 (Application outside Great Britain) Order 1977 No. 1232

The Packaging and Labelling of Dangerous Substances Regulations 1978 No. 209

The Hazardous Substances (Labelling of Road Tankers) Regulations 1978 No. 1702

The Notification of Accidents and Dangerous Occurrences Regulations 1980 No. 804

The Control of Lead at Work Regulations 1980 No. 1248
The Diving Operations at Work Regulations 1981 No. 399
The Dangerous Substances (Conveyance by Road in Road Tankers and Tak Containers) Regulations 1981 No. 1059
The Health and Safety (First Aid) Regulations 1981 No. 917

C. Approved Codes of Practice

Disciplinary Practice and Procedures in Employment. Code of Practice 1977 (ACAS)
Safety Representatives and Safety Committees 1977 (HSC)
Time Off for Training of Safety Representatives 1977 (HSC)
Control of Lead at Work 1980 (HSC)
Health and Safety First-Aid 1981 (HSC)

Appendix III

List of cases referred to in the text

Adsett v K & L Steelfounders & Engineers Ltd [1953] 1 All ER 97
Aitchison v Howard Doris Ltd [1979] SLT (Notes) 22
Alidair v Taylor [1978] IRLR 82
Alphacell Ltd v Woodward [1972] A.C. 824
Anns v Merton B.C. [1978] AC 728
Armour (J.) v Skeen (J.) (Procurator Fiscal Glasgow) [1977] IRLR 310
BAC v Austin [1978] IRLR 332
Belhaven Brewery v McLean [1975] IRLR 370
Boyle v Kodak [1969] 2 All ER 439
Braham v J. Lyons Ltd [1962] 3 All ER 281
Campbell v Wallsend Slipway and Engineering Co Ltd [1978] ICR 1015
Cassidy v Dunlop Rubber Co Ltd [1972] 13 KIR 255
Chant v Aquaboats [1978] ICR 643
Chrysler UK Ltd v McCarthy (J. D.) [1978] ICR 939
Close v Steel Company of Wales [1962] AC 367
Davis (A. C.) & Sons v The Environmental Health Department of Leeds City Council [1976] IRLR 282
Dodd v Central Asbestos Co Ltd [1972] 2 All ER 1135
Dutton v Bognor Regis UDC [1971] 2 All ER 1003
Gannon (J) v Firth (JC) Ltd [1976] IRLR 415
Gardner v Peeks Retail Ltd [1975] IRLR 244
General Cleaning Contractors Ltd v Christmas [1952] 2 All ER 1110
Harrison (TC) (Newcastle under Lyme) Ltd v Ramsey (K) [1976] IRLR 135
ICI v Shatwell [1964] 2 All ER 999
Janata Bank v Ahmed [1981] IRLR 457
Kenna v Stewart Plastics Ltd [1978] HSIB 34 at p 15
Keys v Shoefayre Ltd [1978] IRLR 476
Langston v AEUW [1974] 1 All ER 980
Lister v Romford Ice and Cold Storage Ltd [1957] AC 555

Marshall v Gotham Co Ltd [1954] 1 All ER 937
Mayhew v Anderson (Stoke Newington) Ltd [1978] IRLR 101
McArdle v Andmac Roofing Co [1967] 1 All ER 583
Nico Manufacturing Co Ltd v Hendry [1975] IRLR 225
Nimmo v Alexander Cowan & Sons Ltd [1967] 3 All ER 187
Osborne v Bill Taylor of Hoyton [1982] IRLR 17
Otterburn Mills Ltd v Bulman [1975] IRLR 223
Oxley v Firth [1980] IRLR 135
Pagano v HGS [1976] IRLR 9
Paris v Stepney Borough Council [1951] AC 267
Public Prosecutor v Yuvaraj [1970] 2 WLR 226
Quinn v Burch Brothers (Builders) Ltd [1966] 2 QB 370
R v Swan Hunter Shipbuilders Ltd [1981] ICR 831
Robson v Brims & Co Ltd [1977] HSIB No. 16 at p 16
Secretary of State v ASLEF [1972] 2 All ER 949
Smith v Austin Lifts Ltd [1959] 1 WLR 100
Spencer v Paragon Wallpapers Ltd [1977] ICR 301
St Anne's Board Mill Co Ltd v Brien [1973] ICR 444
Stokes v GKN Ltd [1968] 1 WLR 1776
Tesco Stores Ltd v Edwards [1977] IRLR 120
Sweet v Parsley [1969] 1 All ER 347
Tesco Supermarkets Ltd v Nattras [1971] 2 WLR 1167
Turner v Goldsmith [1891] 1 QB 544
Vacwell Engineering Co Ltd v BDH Chemicals Ltd [1969] 3 All ER 1681
Waugh v British Railways Board [1980] AC 521
Wheat v Lacon (E) & o Ltd [1966] 1 All ER 582
White v Pressed Steel [1980] IRLR 176
Wright v Dunlop Rubber Co Ltd and another [1972] 13 KIR 255

Abbreviations

All ER	All England Law Reports (published by Butterworth & Co, 11/12 Bell Yard, London WC2A 2LG)
ICR	Industrial Court Reports (Incorporated Council of Law Reporting for England and Wales, 3 Stone Buildings, Lincoln's Inn, London WC2A 3XN)
IRLR	Industrial Relations Law Reports (Eclipse Publications Ltd, 286 Kilburn High Road, London, NW6)

ITR	Industrial Tribunal Reports (HMSO, Atlantic House, Holborn Viaduct, London, EC1)
KIR	Knights Industrial Reports (Charles Knight & Co Ltd, 11–12 Bury Street, London EC3A 5AP)
LGR	Local Government Review (Justice of the Peace Ltd, Little London, Chichester, Sussex)
QB	Queen's Bench Division Reports
WLR	Weekly Law Reports (as ICR above)
HSIB	Health and Safety Information Bulletin (Industrial Relations Report and Review) (Industrial Relations Service, 67 Maygrove Road, London, NW6)

Appendix IV

Bibliography

For a more comprehensive list of literature relevant to occupational health and safety see Appendix 14 of the Robens Report and, for Government publications see the Publications Catalogue produced by the Health and Safety Executive, first published in 1980.

Official publications

Report of the Investigation of the Crane Accident at Brent Cross, Hendon on 20 June 1964. HMSO, 1965, Cmnd 2768

Report of the Tribunal Appointed to Inquire into the Disaster at Aberfan on 21 October 1966. HMSO, 1966

Works Safety Committees – some case studies. HMSO, 1968

Report on the Public Inquiry into the Accident at Hixton Level Crossing 1968. HMSO, 1968, Cmnd 3706

Cost effectiveness approach to industrial safety. HMSO, 1972

Noise. Code of Practice for reducing the exposure of employed persons, HMSO, 1972

Report of the Committee on Safety and Health at Work (Robens Report) HMSO, 1972, Cmnd 5034

Report on lead poisonings at the RTZ Smelter Avonmouth, HMSO, 1972, Cmnd 5042

Safety training needs and facilities in one industry, HMSO, 1973

Vinyl Chloride. Code of Practice for Health Precautions. HSE, 1975

Report of the Court of Inquiry into the Flixborough Disaster, 1974. HMSO, 1975

Articles and substances for use at work: guidance notes: general series 8. HMSO

Report on explosion at Houghton Main Colliery, Yorkshire on 12 June 1975. HMSO, 1976

Advice to Employees: guidance notes: HSC Series 5. HMSO

Employers' policy statements: guidance notes: HSC Series 6. HMSO

Advisory Committee on Major Hazards. 1st Report. HMSO, 1976

Report on Explosion at Appleby-Frodingham Steelworks, Scunthorpe, 4 September 1975. HMSO, 1976

Protective Legislation: Who Benefits – Men or Women? Speech of Chairman. EOC 1977

Safety Representatives and Safety Committees: Guidance Notes. HMSO, 1977

Safety Committees: Guidance Notes. HMSO, 1977

Asbestos – Work on Thermal and Acoustic Insulating and Sprayed Coatings First Report of Advisory Committee on Asbestos. HMSO, 1978

Working Conditions in Universities, 1978. HSE Pilot Survey

Working Conditions in Schools and Further Education Establishments, 1978. HSE Pilot Survey

Report on Fire in HMS Glasgow, 23 September 1976. HMSO, 1978

Packaging and Labelling of Dangerous Substances – Guidance Notes. HMSO, 1978

An investigation of potential hazards for operations in the Canvey/Thurrock Area. HMSO, 1978

Advisory Committee on Major Hazards. 2nd Report. HMSO, 1979

Health and Safety Legislation: Should we distinguish between men and women? EOC 1979

Health and Safety, Report on Manufacturing and Service Industries, 1979. HMSO, 1980

Effective Policies for Health and Safety. HSE, 1980

Guide to the Notification of Accidents and Dangerous Occurrences, HS(R) Series 5. HMSO, 1981

Guide to the Health and Safety at Work Act. HS(R) Series 6. HMSO, 1980

Consultative Document on Notification of New Substances. HMSO, 1981

Guidance Notes for the Health and Safety (First Aid) Regulations 1981, HS(R) Series 11. HMSO, 1981

EEC Council Directive on Major Accident Hazards from Certain Industrial Activities. December 1981

Fire and Explosion at Petroflex Ltd, Stoke on Trent, 11 February 1980. HMSO, 1981

Other publications

ATIYAH, P. S., *Accidents Compensation and the Law*, Weidenfeld and Nicholson 1970.

BARRETT, B. N., Safety Representatives, Industrial Relations and Hard Times 6, *Industrial Law Journal*, 1977, 165

— Occupational Safety and the Contract of Employment, *New Law Journal*, 1977, 1011

— Employers' Liability for Work Related Ill-Health 10, *Industrial Law Journal*, 1981, 101

CALABRESI, G., *The Costs of Accidents*, YUP, 1970

CRONER, *Health and Safety at Work*, Croner, Looseleaf Service

HEINRICH, H. W., *Industrial Accident Prevention*, 4th edition, McGraw Hill, 1959

HOWELLS, R. W. L., A New Wave of Interpretations of the Factories Acts, 25, *Modern Law Review*, 1962, 528

— The Robens Report 1, *Industrial Law Journal*, 1972, 185

— Worker Participation in Safety 3, *Industrial Law Journal*, 1974, 87

— Worker Participation in Occupational Safety 4, *Poly Law Review*, 1978, 19

REDGRAVE'S, *Health and Safety in Factories*, Butterworth, 1976

WILLIAMS, J. L., *Accidents and Ill-health at Work*, Staples Press, 1960

INDUSTRIAL RELATIONS SERVICES, *Industrial Relations Review and Report*, bi-monthly journal

Index